Epistemic Ambivalence

This book delves into the complex relationship between religious imaginaries and the perception of space among followers of Candomblé and Pentecostal churches in Belo Horizonte, Brazil's third-largest urban agglomeration.

It adopts a dual perspective, examining the broader political, economic, and social dimensions of these religious communities' urbanisation and spatial distribution and their members' individual beliefs and behaviours. Through this approach, the book aims to provide a nuanced and insider's view of these religious positions, challenging our preconceived notions of urban spaces and contributing to the larger discussion of decolonial urban theory and spatialised post-secular thought.

This transdisciplinary book will appeal to a broad range of researchers, particularly those interested in urban and religious studies. Its strong spatial perspective makes it attractive to architects and urban designers. It will be of interest to those in human geography, urban planning, design, architecture, political science, religious studies, and culture studies.

Daniel Medeiros de Freitas is a professor in the Urban Department at the School of Architecture and Urbanism, Federal University of Minas Gerais (UFMG). He is an author of *Field of Power of the Large Scale Urban Projects* (2017) and one of the editors of the book *City-Estate-Capital: Urban Restructuring and Resistances in Belo Horizonte, Fortaleza and São Paulo* (2018).

Carolina Maria Soares Lima is a Geographer, master in geography, and MBA in social politics. Her research interests include urban arts, the notion of representation and public spaces in Latin America. Carolina is currently a PhD candidate in Architecture and Urbanism at the Federal University of Minas Gerais.

Krzysztof Nawratek is a Senior Lecturer (Associate Professor) in Humanities and Architecture at the University of Sheffield, UK and a Visiting Professor at Pontificia Universidade Catolica de Minas Gerais, Brazil. He is the author of *City as a Political Idea* (2011), *Holes in the Whole: Introduction to Urban Revolutions* (2012), *Radical Inclusivity: Architecture and Urbanism* (ed. 2015), *Urban Re-Industrialisation* (ed. 2017), *Total Urban Mobilisation: Ernst Junger and Postcapitalist City* (2018), and *Kuala Lumpur: Community, Infrastructure and Urban Inclusivity* (co-authored with Marek Kozlowski and Asma Mehan, 2020).

Bernardo Miranda Pataro is an International Analyst and Social Scientist, currently enrolled in the Master's programme in Sociology at the Federal University of Minas Gerais, in Belo Horizonte, Brazil. His research interests include the relationship between religion and violence as well as religion and politics, and production of identities in modernity and international security.

Epistemic Ambivalence

Pentecostalism and Candomblé
in a Brazilian City

**Daniel Medeiros de Freitas,
Carolina Maria Soares Lima,
Krzysztof Nawratek, and
Bernardo Miranda Pataro**

Routledge
Taylor & Francis Group

LONDON AND NEW YORK

First published 2024
by Routledge
4 Park Square, Milton Park, Abingdon, Oxon OX14 4RN

and by Routledge
605 Third Avenue, New York, NY 10158

Routledge is an imprint of the Taylor & Francis Group, an informa business

British Library Cataloguing-in-Publication Data
A catalogue record for this book is available from the British Library

ISBN: 978-1-032-16312-3 (hbk)
ISBN: 978-1-032-17188-3 (pbk)
ISBN: 978-1-003-24801-9 (ebk)

DOI: 10.4324/9781003248019

Typeset in Times New Roman
by KnowledgeWorks Global Ltd.

Contents

Figures and table

Figures

Table

Acknowledgements

We would like to express our heartfelt gratitude to everyone who has contributed to the creation of this book, which represents the culmination of four years of research conducted in Belo Horizonte, Brazil. While it is impossible to name everyone who has supported and encouraged us along the way, we would like to extend special thanks to Gabriela Alcantara Azevedo Cavalcanti de Arruda, whose research has been invaluable to us. We would also like to thank Cynthia Braulio, Roberto Freitas, D. Isabel Casemiro, Keli and Camilo Gan (Baticum Tendinha Cultural), Denise Morado, Thiago Cenettieri, Joao Tonucci, and Bruno Fernandes for their contributions to our work. We are grateful for the institutional support we have received from the UFMG School of Architecture PRAXIS-EA/UFMG and the School of Architecture at the University of Sheffield.

As one of the co-authors of this book, I, Carolina, would like to personally thank my partner, Luana Paris, and my parents, Marcia and Acacio Lima, for their unwavering support throughout the research and writing process. I am also grateful to my teachers and professors, as well as my colleagues and co-authors, for their invaluable insights and contributions. Lastly, I would like to acknowledge Caroline Craveiro, whose guidance helped me better understand the cultural significance of religion and led to my involvement with the PRAXIS group.

Krzysztof would like to extend his gratitude to his wife, Kasia Nawratek, for her unwavering support and patience throughout the four-year process of researching and writing this book.

I, Bernardo Miranda Pataro, one of the co-authors of this book, would like to thank everyone who made this incredible journey possible. I want to thank my partner, Debora Vargas, for her unconditional support over the past four years and her contribution to the research. I also want to thank my family for their support, especially my grandfather, Sebastião Idelfonso de Miranda, who is one of the great intellectual inspirations of my life. A special thanks to my co-authors, without whom none of this would be possible.

1   Introduction

The confusing religious landscape in Brazil

I am evangelic, but I do not like to follow the church's doctrines; I think it's wrong. Each one talks to God in his own way. They say things like, "you should wear skirts, do not cut your hair" I don't agree with that; I don't think that the church should interfere with how you live your life. The church is there to guide you on the path of God, but your intimacy with God has nothing to do with the church.

> (Black woman, 46 years old, living in the Occupied Territory, Belo Horizonte, Brazil, interviewed by Authors in April 2019)

Look, I do not go to church. I was born into a Catholic family; I was Catholic until a certain age. I went to the evangelical church, then I left the evangelical church, I went to the spiritist church, then I left the spiritist church, I went back to the catholic church, and now I do not have any religion. I believe in God, not a religion.

> (Black woman, 53 years old, living in the Occupied Territory, Belo Horizonte, Brazil, interviewed by Authors in April 2019)

Inception. From collaborators to friends

The research began in April 2019 as a small project led by Krzysztof Nawratek focused on the spatial elements of religious imaginaries. During the initial three weeks of fieldwork in the "occupied territories"[1] of Belo Horizonte, it became clear that almost everyone we interviewed belonged to one of many Pentecostal churches. While our interviewees have made us aware of followers of other religions like Candomble and Umbanda living in the area, we have yet to meet them during this research stage. We met a few Catholics, Spiritists, and people with no defined religious affiliation, but Pentecostalism was the dominant religion in the urban outskirts of the city. The global pandemic halted the plans to continue fieldwork in 2020, but the authors decided to continue their collaboration and research. In 2020 and 2021, we conducted interviews using social media (primarily WhatsApp), with only Brazilian team members participating (during the initial investigation in 2019, Krzysztof also participated in interviews and mapping). Fast forward to February of 2022, and Krzysztof Nawratek made the journey to Belo Horizonte again as a visiting professor; his trip was made possible by the generous funding provided by the Brazilian Federal Foundation for Support and Evaluation

DOI: 10.4324/9781003248019-1

of Graduate Education – CAPES (Coordenação de Aperfeiçoamento de Pessoal de Nível Superior).

It's worth mentioning a few words about the authors and research team. As mentioned above, the small pilot project was initiated by Krzysztof Nawratek with a grant from the World University Network. Initial work was supported by the PRAXIS-EA/UFMG research group led by Professors Denise Morado and Daniel Freitas. Interviews were conducted by a group of MA and PhD students. In March 2020, as the world began to grapple with the ramifications of the pandemic, a small group – Gabriela Alcantara Azevedo Cavalcanti de Arruda, Daniel Medeiros de Freitas, Carolina Maria Soares Lima, and Bernardo Miranda Pataro, all hailing from the Federal University of Minas Gerais (UFMG) in Brazil, and Krzysztof Nawratek from the University of Sheffield in the United Kingdom – found themselves brought together meeting regularly – Google Meet linked us together. And yet, despite the distance and the limitations imposed by the ongoing global crisis, the bonds forged between these individuals only grew stronger. This book is co-authored by a group of friends.

Apart from Daniel Freitas and Krzysztof Nawratek, the research team was composed primarily of MA students. Carolina would eventually become a PhD researcher in the School of Architecture at UFMG. The Brazilian team took on the task of conducting interviews, delving deep into the intricacies of the research at hand. At the same time, due to his insufficient Portuguese for effective voice communication via WhatsApp, Krzysztof assumed a more external role, reviewing and contextualising the collected data. This change in how the research was conducted proved enlightening, highlighting the power dynamics between the research team members, researchers, and participants.

As the project progressed, the team realised they were moving towards a flatter, more horizontal structure. The project was co-produced by a group of individuals with different expertise and interests without a leader. None of the team members had formal training in religious studies, but they all shared a keen interest in the topic and some previous experience in interdisciplinary research. And it was through this shared passion and commitment that all of us were able to make meaningful contributions to the project at hand. The methodological aspects of our work will be discussed briefly below and further in Chapter 2.

Almost flat

As the pandemic swept across the globe, many believers turned to the digital realm to continue their religious practices. We followed them there, tracking their movements and documenting their experiences in this new virtual landscape. The data collection process was flattened and made more collaborative, with regular meetings and input from guests outside the research team, including interviewed participants eager to further engage with the project and the research team.

This flattening of hierarchies in the research process was part of our broader ambition to contribute to the ongoing discussion about decolonial research practices. Krzysztof had previously co-authored an article (Nawratek & Mehan, 2020) with similar goals, analysing the dating practices of young Muslims in Malaysia. Carolina and Krzysztof, along with two other academics, contributed to a conversation in the special issue of the *Field* journal on the antiracist and decolonial university (Mehan, Lima, Ng'eno, & Nawratek, 2022). Daniel, Carolina, and Bernardo discussed the first results of the interviews by approaching authors from urban studies and sociology in a publication in the *Revista Indisciplinar* (Freitas, Soares, & Pataro, 2021). And Daniel, through his work with PRAXIS-EA/UFMG,[2] brought our work into the realm of research on power relations and urban production practices, focusing on the reading of place and the use of local narratives in the design of public spaces, oriented towards decrypting access to urban policy.

In this book, we take religion as a lived intellectual tradition in which non-Western ontologies and non-Western epistemologies can be developed. On the one hand, we follow Mircea Eliade's (1959) phenomenological approach to studying religions as something that cannot be reduced to culture, economy, or the study of social relationships. On the other hand, in the context of contemporary discussions on the very nature of religion as an academic, Western construct, we see religion as a particular worldview where all elements of world-making practices are present. The work of David G. Robertson (Robertson, 2021; Valaskivi & Robertson, 2022) on epistemic capital has been particularly influential in shaping our approach. By examining two distinct Brazilian religious traditions (Brazilian Pentecostalism and Candomble) through this perspective, we aim to tell the story of vernacular Brazilian thinking. However, we are fully aware of the potential essentialist bias that can arise when defining what "Brazilian" means. Thus, we offer an incomplete map rather than trying to create a universal model.

We attempt to take a clear decolonial stance in this book by avoiding, as much as possible, referencing "dead white males" and instead using as many Brazilian or South American references as possible. However, we cannot avoid referring to thinkers who are essential and helpful in building our argument (Eliade, Bordieu, Lefevbre, etc.). Still, we try to refer to them through their work's "local", South American reception. In Chapter 4, we will develop this discussion further and attempt to conceptualise most of our findings.

(De)politicising religions

In Chapter 3, we delve into the intricacies of religious activities in Brazil, examining two distinct aspects that we have initially chosen to label as "political" and "missionary". However, this classification is not meant to be taken at face value, as engagement (or lack thereof) in politics is merely a "byproduct" of more fundamental theological strategies.

By "political" (or "politics-focused"), we referred to a set of strategies that aim to establish and expand the influence of religion into the non-religious sphere, such as urban space, politics, institutions, and media, to help believers grow spiritually. This approach could be seen as an example of Dominion Theology (Burity, 2021; Garrard, 2020); however, we also use this framework to discuss non-Christian religions in Brazil.

On the other hand, by "missionary" (or "missionary-focused"), we referred to strategies aimed at strengthening the spiritual power of individual believers. While the political approach seeks to create "religious infrastructure" as an external "prosthesis" for individual believers, the missionary approach focuses on preparing individual believers to face the perceived hostilities of the secular world. At the end of the day, we decided to rename these categories into "inward" and "outward" looking, as the Reader will see in the following chapter.

Fading Catholicism, growing Pentecostalism

As one might expect, Catholicism is still the biggest religion in Brazil, with institutions and broad support from the media and social networks. A legacy of Portuguese colonisation, Catholicism was the official state religion until the Republican Constitution of 1891, which instituted the secular state. The majority of the Brazilian population is Christian (87%), with the majority of them being Roman Catholic (64.4% in 2010, though this has decreased from 95% in the 1970s). We believe it will be around 50% in 2022. However, we lack the recent data.

In recent decades, one of the defining characteristics of Brazilian Catholicism has been the plurality of discourses (Mariz, 2006) and the vast distance between liberation theory and charismatic renewal tendencies. The former critiques the Eurocentric and peripheral disconnection of modern theology, advocating for an inductive method that does not start from theological interpretations of reality but rather from the reality of poverty, exclusion, and liberation to make theological reflection and invite transformative action. On the other hand, the latter is an effort to bring the faithful closer together, guided by greater (assumed) similarity to the early church and faith healing, as well as more intense use of media and strategies to increase the number of followers. In Brazil, this tendency was influenced by the global trends in the Catholic Church and North American Pentecostalism in the 1960s and 1970s and, more recently, by Brazilian (Neo)Pentecostal strategies. According to Carranza (2011), the charismatic strategy adopted by the Catholic Church has not proven to be an effective response to the advance of Pentecostalism. The more it appropriates media strategies, the more it resembles (Neo)Pentecostalism. The liberation theory has also been reduced in scope and has yet to attract new followers, particularly among younger generations. What these two tendencies have in common, which is also crucial in Pentecostalism, is a focus on believers' individual or shared experiences. We argue (following Mariz, 1992 and de Almeida & Barbosa, 2019) that the evolution of Brazilian Catholicism in the late 20th century was both a reaction to the growing competition from Pentecostalism and a preparation for the increasing diversity of beliefs in Brazil.

In a political sense, Catholicism still has a significant presence in practically all Brazilian cities. Old and newly erected buildings hosting various catholic organisations and temples still dominate the political and economic landscape, exerting their influence on culture. But as the Catholic Church tries to keep its position, it faces a conundrum. How to reconcile the complexities of its doctrine with the demands of a rapidly evolving society? The call for a return to traditional practices clashes with those who advocate for a Church more attuned to the plight of the marginalised and poor. And at the centre of it is how to provide salvation to those in need in a constantly shifting and changing world.

Two key moments in the history of Belo Horizonte illustrate the powerful influence of the Catholic Church in shaping the city's urban landscape. In the original plans for the city, the cathedral was meant to play a prominent symbolic role, situated at the end of Avenida Afonso Pena, the main linear axis of the city, opposite the central market. But in the early stages of construction, the Church opposed this proposal and insisted on preserving the original Church of Curral Del Rei. This small village had been demolished to make way for the new city in a more central location. Less than two decades later, they again pressed to construct a new cathedral, the current Cathedral of Boa Viagem.

A second example, four decades after the inauguration of the original project, the city boasts a critical modern architectural complex, the Pampulha Modern Ensemble, built around an artificial lake, Pampulha Lake. It includes a casino, a ballroom, the Golf Yacht Club, and the Church of Saint Francis of Assisi. Designed by Oscar Niemeyer and featuring a mural by Candido Portinari, the exterior landscape was designed by Roberto Burle Marx. In 2016, the building was classified as a World Heritage Site, reinforcing the importance of this site as a tourist destination. These examples demonstrate the Catholic Church's presence in the city and its impact on the urban landscape. Other religious landmarks, such as the "Praça do Papa" (Pope's Square), where the first visit of a pope to the city took place in 1980, and the "Monumento à Paz" (Monument to Peace) built in 1983 also serve as reminders of the Church's influence, the symbolism of these structures, such as the cross, the sculpture, and the square, all point to the importance of Christianity in the city.

Christianity in Brazil is caught in a paradox. On the one hand, it espouses salvationist ethics and grand, universal principles, but on the other, there is mounting pressure to focus on the people's immediate needs. This has led to a migration of believers from traditional Protestantism and Catholicism towards Pentecostalism and even atheism (Mendonça, 2006). Regarding political engagement, traditional Protestantism in Brazil is influenced by the liberal and republican ideals of North American missions and a modern, pragmatic approach to education. However, in Brazil, this liberal project lacks a unifying vision that inspires action worldwide, resulting in a more individualistic and cultured religiosity within communities. Even the most conservative Evangelical churches, when oriented towards evangelism, tend to focus on forming small, Bible-centred communities rather than mass religion. Their political representatives are limited to corporatist practice rather than a broader power project.

With the rise of Pentecostalism, a fundamental shift began to occur. Organised in three waves of expansion (Freston, 1993)[3] Pentecostalism has seen tremendous growth in Latin America and parts of Africa, spreading through cross-cultural missions and faith-based outreach, particularly among the lower-middle and poor classes of major urban centres (Mariano, 1999).

Is there any religion one cannot find in Brazil?

Before delving into the specifics of Brazilian Pentecostalism and Candomble, it is worth taking a step back to examine the broader religious landscape in the country. However, it is essential to note that while we will attempt to discuss all of the most important religious or spiritual positions, we cannot cover the entire spectrum.[4]

We will be discussing the following religions briefly: Kardecist (Spiritism), "African Matrix" (Candomble and Umbanda, but also mentioning Congado, even though it is a part of Catholicism), and Neo-esoterism. Additionally, we will briefly mention Islam, Judaism, Buddhism, Shinto, and new Japanese Religions, as the largest population of Japanese origin living outside of Japan are in Brazil. We intend to draw attention to the characteristics that best contextualise each religion as it consolidated in Brazil.

Kardecism, in conjunction with Catholicism and Evangelical-Pentecostal churches, holds significant sway in Brazil, not for its number of followers but for its dissemination of symbolic references and prestige among the educated urban middle classes. This leads to broad exposure in the media, prime locations for Kardecist temples in cities, and a prominent presence in the publishing industry.

Considered an avant-garde religion in the country, Kardecism's most significant attraction lies in its unique blend of "experimental science" and revealed faith (Lewgoy, 2011). Heavily influenced by its French heritage, Brazilian Spiritism emphasises reason and free will, shaping a reflective and internalised religious practice, but with an essential supernatural component that, as we shall see, brings elements of Christianity and African-based religions together. A large part of its acceptance in Brazil comes from its ability to reconcile rational and religious discourse, making it more palatable to social layers that condemn "primitive" or "superficial" religions. In Kardecism, the idea of a distant God is circumvented by spiritual guides who, through mediumship, help individuals to evolve through successive incarnations, free will, and learning from life on earth.

In terms of political engagement, Brazilian Spiritism has a slow diffusion and detraditionalisation of its ways of believing and belonging. The educated frontier operates as a barrier to the inclusion of low-educated adherents, forming a group with erudite values and strong Christian legitimacy of mediumistic practices. This legitimation played a decisive role in building Umbanda, a Brazilian religion that started from the political and missionary combination between Christianity and elements of African religions.

In Brazil, African-based religions are largely uninstitutionalised, with doctrines and practices passed down through oral tradition in autonomous "terreiros". These practices are often characterised by trance, the worship of spirits, and rituals frequently dismissed by dominant religions as primitive and demonic possession, divination, sacrifices, and other magical practices. Despite this, they hold high cultural legitimacy and a solid connection to popular festivals, carnival, black movements, music, and literature, which contributes to their widespread diffusion throughout the country, each with its own regional characteristics. One particularly intriguing example is Congado, which combines elements of African traditions within the institutionalised framework of the Catholic Church.

The complexity of African-based religions in Brazil is a prime example of why the term "religion" is problematic. These beliefs are deeply ingrained in Brazilian culture and practices, making it almost impossible to separate culture, black heritage, and spirituality. Take Candomblé, for instance. It arrived in Brazil with a vital ethnic component, organised into small groups known as "families of saints", and through cultural elements such as music, dance, oral history, and popular traditions. Over time, these elements have incorporated Catholic and European influences, diluting ethnic relations by merging the worship space of the terreiros and through the initiation process where the faithful assume a religious name and commit to Orixa and the "father/mother of saints". Today, Brazilian Candomblé can be divided into three main groups: the Jêje-Nagô nation and Queto nation, both of which are closer to its African origins, and the Angola nation, which is more open to indigenous and Catholic influences.

The brutal suppression of the Candomblé religion, viewed with suspicion as a means of organising enslaved black individuals, was one of the critical limitations on its political structure. For instance, displaying symbols on the buildings used for worship was banned, and the terreiro, a sacred space with restricted access, serves as both a dwelling for the gods and a place of worship and gathering. We will delve deeper into the intricacies of Candomblé in the following chapters of the book.

On the other hand, Umbanda emerged from the romantic idealisation of indigenous culture and the elevation of certain popular elements to the status of national culture. This was part of a movement that tried to apologise for racist prejudices and practices that characterised Brazilian culture in the 20s and 30s. At that time, there was a surge in the study of Afro-Brazilian religions and significant interest from intellectuals who sought to reflect their vision of the groups that make up Brazilian society through the valorisation of popular culture and black values. As previously mentioned, Umbanda blends elements of Kardecism, Catholicism, and Afro-Brazilian traditions. It was created with the sociocultural mission of uniting the races and classes that make up Brazilian society. It is an intermediate and conciliatory religious form. For example, Umbanda practices incorporate the concept of karma, spiritual evolution, and communication with spirits and reduce the intensity of stigmatised elements such as animal sacrifice, frenzied dancing, and alcohol, tobacco, and gunpowder. In the political dimension, Umbanda has a robust organisational structure inspired by civil society, distinct from the mythical families and kinship relationships of Candomblé.

Islam is a minority religion in Brazil, with less than 40,000 believers. The number of Muslims in Belo Horizonte is particularly small, numbering fewer than 1,000. Still, as Cristina de Castro and Igor Gonçalves Caixeta (2021) have noted in their recent research, it is an influential and growing community. In the context of this book, it is worth mentioning this religion for two main reasons. Firstly, there is the historical trajectory of Brazilian Islam (Oliveira, 2006). As a colonial country of immigrants, Muslims in Brazil arrived as slaves or, later, as immigrants from the Middle East, mostly from what is known as "Great Syria". Some Muslim scholars argue that the history of Islam in Brazil parallels Christianity, as there were converts to Catholicism among the Muslim population during the first wave of colonisation. Around 20,000 black Muslims lived in the country in the 19th century (mainly in the region of Salvador, but also in Rio de Janeiro and Sao Paulo), but these communities mostly disappeared by the mid-20th century, with reports of converts to Umbanda or Catholicism. The immigrants from the late 19th century also make for an interesting case – Christians and Muslims were coming from the same part of the world. Still, both communities vigorously protected their identities and practices, resulting in self-imposed distance from one another in Brazil. The third reason why mentioning Islam is relevant in our work is the current profile of the Muslim community in the country. This group's level of education and income places them among the top strata of Brazilian society – contemporary Islam in Brazil is associated with the intellectual elite. One of the most influential Brazilian theorists of the far right, Olavo de Carvalho, has been, for some time, a Muslim himself (Sedgwick, 2020).

Like many other religions in the country, Brazilian Judaism displays a unique hybridisation of beliefs and practices. The 2010 census reported 107,329 individuals identifying as Jewish, which has grown from 90,000 in 1980. However, these statistics are problematic as they only consider self-identification and do not account for non-religious Jews or groups claiming to be descendants of Marranos, who were forced to convert to Catholicism during the Inquisition and now define themselves as religious Jews. In recent decades, there has been a proliferation of Messianic synagogues and other groups seeking legitimisation as Jews by Orthodox rabbis in Brazil and abroad. The topic of "Christian Zionism" (Carpenedo, 2021; Machado, Mariz, & Carranza, 2022) and its relationship to far-right politics and the Bolsonaro administration has further complicated the already nuanced understanding of Brazilian Judaism and its place in the political landscape.

Finally, discussing ethnic-connected religions, we must mention Shinto, Buddhism, and new Japanese religions. As mentioned, around 1.5 million Brazilians have Japanese heritage, approximately 100,000 Chinese people, and some Koreans live in Brazil. These groups contribute to the presence of Buddhism in Brazil. There is some interest in Buddhism among Brazilians without any Asian heritage, but it is relatively small and mostly confined to urban intellectual elites. Shintoism remains mainly exclusive to the Japanese community, and so-called "new Japanese religions" have had slightly more success reaching beyond this ethnic core but are still relatively minor. In census classification, these religions are grouped under the general category of "other oriental religions". Additionally, there is a broader, albeit the small, group of religions commonly referred to as "neo-esoteric" (Magnani, 2006), which includes Eastern religions, indigenous cosmologies, European esoterism, North American transcendentalism, and even "science" (among followers of the New Age movement, for example). These religions maintain a transnational character and spatiality in Brazil similar to those identified globally, alternating between small places in large urban centres and more extensive facilities outside cities.

From one religion to many worldviews

As we begin our examination of the shifting religious landscape of Brazilian cities, it bears mentioning that our focus will primarily be on Candomble and Pentecostalism. However, as with any such inquiry, references to the broader religious milieu, especially the position of the Catholic Church, will inevitably find their way into our discourse. The past 40 years have seen a tectonic shift in the religious landscape of Brazilian cities, with a marked decline in the number of Catholics and a corresponding explosion of new Pentecostal churches. One need not look far to see the effects of this change – the proliferation of these churches across the urban landscape is a testament to the rapid increase in Pentecostal believers. It is worth noting that this shift is not a recent phenomenon. In fact, the number of Catholics in Brazil has been steadily decreasing since 1970, from a high of 95% to a current figure of 65%. Meanwhile, the number of Protestants, particularly Pentecostals, has been on the rise, with a corresponding increase from 5% to 22%. And it is telling that more than 60% of these Protestants belong to one of the many Pentecostal churches that have emerged in recent years. This transformation of the religious landscape of Brazilian cities is a complex and multifaceted phenomenon, but as we delve deeper into this inquiry, it is essential to keep in mind the sheer scale of this shift – the impressive number of new Pentecostal churches erected in recent years is a testament to the power of this movement and its growing influence on Brazilian society.

What we are witnessing in the religious landscape of Brazilian cities is nothing less than replacing one faith with many others. And make no mistake; this is not a secularisation process but a shift in the fabric of religious belief and practice. This phenomenon is evident not only in the number of believers but also spatially, socially, and culturally. The growth of Pentecostalism is a phenomenon with strong spatial and social connotations. Pentecostalism is a religion of the "bottom half" and "upper ten" per cent of Brazilian society, while the squeezed middle remains predominantly Catholic.

It is important to note that while Pentecostalism is a global religious movement, Brazilian Pentecostalism differs from other forms, such as Nigerian or North American churches. Some researchers claim that Pentecostal churches thrive in places mainly unaffected by the Western enlightenment (Kay, 2013), while others argue that Pentecostal churches are a powerful decolonising force (Low, 2020), "purifying" Christianity (in Brazil mainly in the form of the Catholic Church) from "Western contaminations". As the Pentecostal movement becomes increasingly diverse, we can observe a transformation from the relatively harmonious (concerning cultural and religious imaginaries) Catholic society to a society with a plurality of imaginaries. It is worth noting that this perceived harmony has been achieved through the persecution of African religions and the "secretly syncretic" nature of Brazilian Catholicism. But even if this perceived "harmony" was built upon structural oppression and lies, the contemporary religious landscape in Brazil is significantly more violent. The Pentecostal movement has become increasingly hostile towards the Catholic Church and African Religions, leading us to believe that by studying Pentecostal churches, we can also understand the contours of "other religions" in the country. The edges between Pentecostals and "the others" may reveal much about the conflicts and tensions in Brazilian society.

In this vein, we have focused on Candomble as the primary representative of "other" religions. With roots in the beliefs of Yoruba people brought to Brazil as slaves, Candomble has long been persecuted and forced to hide in syncretic versions of Catholicism. However, as a Brazilian religion constructed from various beliefs of African descendants, Candomble is syncretic in its own right. It is also fundamentally different from Christianity in its focus on oral tradition and rituals (there is no "sacred book", no equivalent of the Bible) and in its very different ethics (there is no equivalent of "ten commandments").

It's worth noting that Pentecostal movements are a rapidly growing denomination of diverse churches. While particular churches differ significantly in doctrinal details, they are characterised by the intense socio-economic mobilisation of their members. Some Pentecostal churches are highly influential politically, with strong economic positions and diverse demographics; some cut across social classes, while others have a homogenous socio-economic profile. Many Pentecostal churches of different denominations are spatially located in proximity to each other, leading us to see this phenomenon as a kind of "church cluster".

All of this serves to underline the complexity of the religious landscape in Brazil and the unique opportunity it presents for exploring the intersection of architecture, urbanism, and religious studies.

Our book explores the intersection of religion and the urban landscape on two distinct scales. On the one hand, we examine the distribution of religious practices within the city in the context of the socio-economic divisions that shape Brazilian society. On the other hand, we focus on the individual experiences of believers, seeking to understand how a particular set of beliefs informs their perceptions of the urban spaces they inhabit and their spatial practices.

It is worth noting that our inquiry is also site-specific, focusing on the Brazilian city of Belo Horizonte. As the capital of the state of Minas Gerais in the southeast of Brazil, Belo Horizonte is home to a population of around 2.5 million, making it the sixth largest city in the country. Belo Horizonte was one of Brazil's first planned regional capitals, with its origins in 19th-century European urbanism, specifically the work of Ildefonso Cerdá and the Polytechnic School of Paris. Its planning and construction took place at the end of the 1890s under the direction of Aarão Reis and Francisco Bicalho. This rich history and unique urban landscape make Belo Horizonte an ideal location for examining how religion and the urban environment intersect and inform one another. Belo Horizonte has also hosted major international sporting events, such as the 2014 FIFA Confederations Cup and the 2016 Summer Olympics. In preparation for these mega-events, significant architectural and urban projects were undertaken to accommodate the influx of visitors and athletes. These works have drastically altered the city's landscape and how space is produced and utilised. One such example is the renovation of the Mineirão stadium, which now often hosts the traditional Catholic event "Torcida de Deus", organised by the Archdiocese of Belo Horizonte. Held annually since 1975,

Torcida de Deus gathers thousands of faithful at the Mineirão stadium to publicly profess their faith in the Blessed Sacrament. This event serves as yet another reminder of the strong presence of the Catholic Church and its followers in the city. This presence can be traced back to the colonisation of Brazil, led by the Catholic Church through the catechesis of the native population and the construction of churches in all Brazilian cities.

As we discuss in Chapter 5, Belo Horizonte is a highly complex and unequal urban entity, characterised by stark disparities in education and income. The city has developed over time, from its 19th-century "European" core to more recent "occupied territories". To some extent, the distribution of religious buildings (churches, temples, and terreiros) is interconnected with these socio-economic conditions. Still, the most apparent spatial feature is historical – Catholic churches are more prevalent in the older parts of the city. In contrast, a much larger number of Pentecostal churches can be found in the newer districts. But as we argue in Chapter 6, the spatiality of religious practices in the city can only be fully understood by delving deeper into the scale of the house and the street, examining the practices of individual believers. In Chapter 8, we explore the realm of online religious groups and the insights gained during the pandemic. And in Chapter 9, we bring the book to a close by returning to the central theme of decolonising our understanding of religious spaces.

Remarks and disclaimers

We hope that this introductory chapter has effectively conveyed the religious landscape in Brazil and our intentions and objectives for this book. However, there are two more aspects that we would like to bring to the Reader's attention. First, we would like to clarify the usage of the term Pentecostalism. As we will discuss further in the book, there are similarities and convergencies between Pentecostal, Neo-Pentecostal, and some other Protestant churches in a charismatic tradition. We decided to put the label "Pentecostalism" on them because they have more similarities than differences. Moreover, we know that in Brazil, the term Pentecostalism is used relatively rarely, with most of the people we have interviewed calling themselves "evangelicals". However, using the word "evangelical" would probably confuse Readers even more; therefore, we stick to Pentecostals.

This book is a collaborative effort with four co-authors, and none of us has English as our first language. Additionally, all of our interviews were conducted in Brazilian Portuguese. We have made a concerted effort to be clear and precise in our translations from Portuguese to English while preserving our research participants' unique idioms and speech patterns. This may result in certain sections of the book sounding uncomfortable for native English speakers. In terms of the narrative structure, the book is intended to be fluid, with moments where we speak in a unified voice, share our individual perspectives without identifying the narrator, and rare instances where one of us wants to have their voice clearly distinguished. We understand that this may be challenging for Readers, but we believe that this linguistic diversity adds depth and nuance to the book and ultimately enriches the Reader's understanding of the religious landscape in Brazil.

Finally, a few words to describe the process of writing this book. It has been a collaborative and dynamic one. As a team, we utilised Google Docs as our central hub for collaboration, with each author starting by contributing two chapters to the document. From there, we engaged in a thorough cross-review and rewrote the entire book, with each author working on chapters initially written by their colleagues. Our efforts to perfect the text were supported by the Grammarly.app, which helped improve the text's grammar and style.

Notes

1 Following Ermínia (2003), we define the term "occupied territories" or "urban occupation" in Brazil in the context of the action of social movements and their pressure on the government to implement housing policies. The term occupation, which differs from the term invasion, recognises the legitimacy of giving utility to an unused space that does not fulfil its social function. Another important distinction occurs between the term urban occupation (planned and structured action with discourses and purposes that go beyond the housing issue and seek a broader political opposition) and the term favela (urban space that results from spontaneous and gradual processes of land occupation).

2 Funded in 2009, PRAXIS-EA/UFMG is a CNPq research group hosted by the Department of Projects and the Postgraduate Program in Architecture and Urbanism (NPGAU) of the UFMG School of Architecture, with projects supported by Fapemig, CNPq, Capes, PRPq and ProEx/UFMG, Ford Foundation and FUSP. It is coordinated by Professors Denise Morado Nascimento and Daniel Medeiros de Freitas. The group is part of the Research Group Program of the Institute for Advanced Transdisciplinary Studies (IEAT/UFMG).

3 The first wave includes Classical Pentecostalism, represented by the *Congregação Cristã no Brasil* (1910) and *Assembléia de Deus* (1911); the second wave consists of the *Igreja do Evangelho Quadrangular* (1953), *Igreja Brasil para Cristo* (1955), *Igreja Deus é Amor* (1962), and *Casa da Benção* (1964); and the third wave includes *Igreja Nova Vida* (1960), *Igreja Universal do Reino de Deus* (1977), *Igreja Internacional da Graça de Deus* (1980), *Igreja Cristo Vive* (1986), *Comunidade Evangélica Nossa Terra* (1976), *Comunidade da Graça* (1979), *Igreja Renascer em Cristo* (1986), e *Igreja Mundial do Poder em Deus* (1998).

4 To understand better the rich and complicated religious landscapa in Brazil, we would advice to read *Handbook of contemporary religions in Brazil*, by Bettina Schmidt and Steven Engler (2016).

References

Burity, J. (2021). The Brazilian conservative wave, the Bolsonaro administration, and religious actors. *Brazilian Political Science Review*, *15*. https://www.scielo.br/j/bpsr/a/K6W Pj8yxktVRMQcqcxpWQFc/?lang=en

Carpenedo, M. (2021). Christian Zionist religiouscapes in Brazil: Understanding Judaizing practices and Zionist inclinations in Brazilian Charismatic Evangelicalism. *Social Compass*, *68*(2), 204–217.

Carranza, B. (2011). *Catolicismo midiático*. São Paulo: Editora Idéias & Letras.

Castro, C. M., Caixeta, I. G. (2021). Islamic practices in Belo Horizonte: Adaptations and choices in a bastion of Brazilian traditionalism. *Social Compass, 68*(2), 190–203.

Cavalcanti de Arruda, G. A. A., Freitas, D. M., Soares Lima, C. M., Nawratek, K., & Miranda Pataro, B. (2022). The production of knowledge through religious and social media infrastructure: World making practices among Brazilian Pentecostals. *Popular Communication, 20*(3), 208–221.

de Almeida, R. R. M., & Barbosa, R. J. (2019). Religious transition in Brazil. In *Paths of inequality in Brazil* (pp. 257–284). Cham: Springer.

Eliade, M. (1959). *The sacred and the profane: The nature of religion* (Vol. 81). Boston, MA: Houghton Mifflin Harcourt.

Ermínia, M. (2003). Metrópole, legislação e desigualdade. *Estudos avançados, 17*, 151–166.

Freitas, D.M., Soares, C. M., & Pataro, B.M. (2021). Idealização do mundo e leitura do lugar nos espaços de religiosidade: entrevistas realizadas nos territórios populares de Belo Horizonte. *Revista Indisciplinar, 7*(2), 124–149.

Freston, P. (1993). *Protestantes e política no Brasil: da Constituinte ao Impeachment, 303s* (Doctoral dissertation, Tese de Doutorado em Sociologia. Departamento de Ciências Sociais no Instituto de Filosofia e Ciências Humanas da Universidade de Campinas).

Garrard, V. (2020). Hidden in plain sight: Dominion theology, spiritual warfare, and violence in Latin America. *Religions, 11*(12), 648.

Kay, W. (2013) Empirical and historical perspectives on the growth of Pentecostal-style churches in Malaysia, Singapore and Hong Kong. *Journal of Beliefs & Values, 34*(1), 14–25.

Lewgoy, B. (2011). Uma religião em trânsito: o papel das lideranças brasileiras na formação de redes espíritas transnacionais. *Ciencias sociales y religión. Porto Alegre, RS, 13*(14) (set. 2011), 93–117.

Lourenço, T. C. B. (2017). Ocupações urbanas em Belo Horizonte: conceitos e evidências das origens de um movimento social urbano. *Cadernos de Arquitetura e Urbanismo, 24*(35), 182–217.

Low, U. W. (2020). Towards a pentecostal, postcolonial reading of the New Testament. *Journal of Pentecostal Theology, 29*(2), 229–243.

Machado, M. D. D. C., Mariz, C. L., & Carranza, B. (2022). Genealogy of the Evangelical Zionism in Brazil. *Religião & Sociedade, 42*, 225–248.

Magnani, JGC (2006). The neo-esoteric circuit. In *Religions in Brazil: continuities and ruptures*. Petropolis: Voices.

Mariano, R. (1999). *Neopentecostais: sociologia do novo pentecostalismo no Brasil*. São Paulo: Edições Loyola.

Mariz, C. L. (1992). Religion and poverty in Brazil: A comparison of Catholic and Pentecostal communities. *SA. Sociological Analysis, 53*, S63–S70.

Mariz, C. L. (2006). Catolicismo no Brasil contemporâneo: reavivamento e diversidade. *As religiões no Brasil: continuidades e rupturas*. Petrópolis: Vozes, 6.

Mehan, A., Lima, C., Ng'eno, F., & Nawratek, K. (2022). Questioning hegemony within white academia. *Field, 8*(1), 47–60.

Mendonça, A. G. (2006). *Evangélicos e pentecostais: um campo religioso em ebulição. As religiões no Brasil: continuidades e rupturas* (pp. 98–110). Petrópolis: Vozes.

Nawratek, K., & Mehan, A. (2020). De-colonizing public spaces in Malaysia: Dating in Kuala Lumpur. *Cultural Geographies, 27*(4), 615–629.

Oliveira, V. P. D. (2006). Islam in Brazil or the Islam of Brazil?. *Religião & Sociedade, 26*(1), 83–114.

Robertson, D. G. (2021). Legitimizing claims of special knowledge: Towards an epistemic turn in religious studies. *Temenos*, *57*(1), 17–34.

Schmidt, B., & Engler, S. (Eds.). (2016). *Handbook of contemporary religions in Brazil*. Leiden: Brill.

Sedgwick, M. (2020). Traditionalism in Brazil: Sufism, Ta'i Chi, and Olavo de Carvalho. *Aries*, *21*(2), 159–184.

Valaskivi, K., & Robertson, D. G. (2022). Introduction: Epistemic contestations in the hybrid media environment. *Popular Communication*, *20*(3), 153–161.

2 Flattening power structure (we have been being preached to)

On October 12, 1492, the natives discovered that they were Indigenous, discovered that they lived in America, discovered that they were naked, discovered that sin existed, discovered that they owed allegiance to a king and queen of another world and a god of another heaven. And that this god had invented guilt and clothing and had burned alive anyone who worshipped the sun, the moon, the earth, and the rain that wets it. They came. They had the Bible, and we had the land. And they told us: "Close your eyes and pray". And when we opened our eyes, they had the land, and we had the Bible.

(Galeano, Eduardo. Los hijos de los días. Madrid: Spain, Siglo XXI de España Editores, 2012)

Now that you came here, and you're talking with us, seeing us, what will you do? Not for you, but for us.

(Queen of Congado, Concórdia Neighbourhood, after a field visit)

Formation

Eduardo Galeano[1]'s excerpt highlights the centrality of religion in the formation of Brazilian society, as well as the ways in which the legacy of colonisation continues to shape contemporary knowledge production in the Global South. From the indigenous worship of sky and river gods to the forced conversion to Christianity, religion has played a fundamental role in shaping the social, political, and economic systems of the Americas.

The arrival of the Europeans in Brazil brought with it the forced conversion of the indigenous population to Christianity and all the philosophical baggage that came with it. This process has had a lasting impact on Brazilian society, as the exclusionary system established during colonisation remains rooted in the country's labour division. Even though Brazil is a very mixed society, light-skinned Brazilians who are often perceived as Latin when abroad have the privilege of passing as white within the country, this privilege allows them, for example, to obtain a higher education. Nearly 83% of post-graduate students in Brazil are white.

As mentioned before, we recognise religion as a form of knowledge, a worldview, and a set of rules and values. Colonialism has imposed certain sets of values

DOI: 10.4324/9781003248019-2

and accepted ways of thinking on the colonised groups, not only in terms of Catholic religiosity but also in European thought and knowledge production. This process has excluded and silenced groups producing knowledge based on other sources. Galeano and Borges (1993) once said, "The Church says the body is a fault, science says it's a machine, advertising says it's business, and the body says I am a party". The body is a rich source of experience, both religious and mundane, and is, as discussed further in Chapter 6, a source of knowledge. As such, we want to consider the sources of knowledge of all subjects as equals if we hope to break the colonial violence cycle that ignores marginalised people's epistemes. Our project is transdisciplinary (Durose, Perry, & Richardson, 2021; Harloe & Perry, 2009; Perry & May, 2010). It means that we recognise not only different academic traditions but also go beyond the university, inviting participants in this research to help us conceptualise its findings (the process culminated in discussions with participants about our preliminary findings and conclusions, the discussions we present in the final chapter).

The inquiry that gave birth to this book was spawned by Krzysztof Nawratek's curiosity about the ways in which individuals of different religious persuasions might experience the same physical spaces. The question at the heart of this investigation, though seemingly simple, was a profound one: if three people of different faiths were to enter a shared space – say, a shopping mall – would they all perceive it in the same manner? And, more importantly, would their religious beliefs influence their perception of that space in any way? To answer this question, Krzysztof joined on a journey, conducting interviews with individuals of various religious backgrounds in Sheffield, UK. But the initial findings were perplexing – the primary distinction that emerged was based on gender, not religion. Regardless of their faith (be it Christianity, Islam, Buddhism, or Umbanda), women tended to value open, green spaces, viewing them as environments that facilitated a connection with the divine. Men, on the other hand, seemed to place more emphasis on temples and rituals. These unexpected results prompted Krzysztof to expand the scope of his research beyond the European context. And so, Krzysztof reached out to the PRAXIS research group at the Federal University of Minas Gerais, with whom he had previously collaborated on another project, and journeyed to the bustling city of Belo Horizonte in 2019 to continue his inquiry in a new cultural context. But as 2020 dawned, a global pandemic descended upon Brazil and the world, bringing a host of new concerns and challenges related to faith and religion. In response, we decided to conduct a new round of interviews remotely to understand how religious beliefs might shape perceptions of space amidst the unique experiences of the pandemic (a topic we will delve deeper into in Chapter 7). These interviews continued until 2021 when in-person meetings became possible again, and we focused on one specific neighbourhood in Belo Horizonte: Concórdia. This book presents this four-year investigation's findings, which spanned from Sheffield to Belo Horizonte.

The religions under investigation proved to be as mercurial as the scales of the analysis themselves, shifting and evolving as our research progressed. Take Sheffield, for example, where Krzysztof conducted interviews with a diverse array of believers without any preconceived notions of which faith traditions would be represented. But upon arriving in Belo Horizonte and delving into the beliefs of the urban periphery, a striking presence of Evangelicals emerged, prompting us to hone in on this particular group. Of course, not all residents of the periphery subscribed to Pentecostalism – we also made a point to speak with followers of other religions, including those with African roots. During these interviews, the tensions between Evangelicals, particularly members of IURD,[2] and practitioners of Afro-Brazilian religions like Candomble became sharper. And so, with a sense of curiosity and trepidation, we decided to discuss these two religious movements – Pentecostalism and Candomble – in this book.

Parallax

In a world where the echoes of colonialism still reverberate, shaping the distribution of power and privilege in insidious ways, it is incumbent upon those privileged enough to hold sway in academia to actively work towards decolonising our research and writing. As authors, we occupy a position of relative authority within the academy, and with that comes a certain level of respect that extends beyond the university walls. It is precisely because of this that we feel a sense of responsibility to amplify the voices of the oppressed and marginalised groups who have been rendered invisible by the forces of coloniality. We know that in Brazil, access to higher education is a privilege we carry with us as we navigate these issues. We're guided by Chakravorty Spivak's "Can the subaltern speak?" (2015), a seminal essay that lays out how developing countries' perspectives are represented by Western discourse and how that prevents marginalisation groups from speaking for themselves. This is especially relevant when considering the inhabitants of Brazilian cities' peripheries.

This book is a collective effort, with the voices of all the authors and some of the participants woven together in a third-person narrative. But there are moments when we, as individuals, want to speak in our own singular voice. Like me, Carolina: The truth hit me like a lightning bolt in a moment of clarity, back when I was just a young woman conversing with a Candomblecista. My previous references came from a white Catholic background, where I learned about the sacred and the profane, good and evil. But it never occurred to me that everything in the world could be considered sacred as the Candomblecista did. This revelation made me rethink my own perception of the world and how it differed from theirs. After that conversation, I saw the world through the Candomblecista's eyes and realised that I hadn't just interviewed him but had been educated and changed by him. And so what I'm writing now is possible because of that experience.

During one of his early field visits, another author, Daniel, encountered a familiar issue: the struggle to transcend the power disparity between the white male interviewer and the interviewee. This consistently impacted the interviewee's posture and responses in some manner. The author recalls a transformative moment in the interview when asked about the significance of faith in one's life. An older black woman hesitated before saying, "Do you truly desire to grasp the essence of faith?

I will endeavour to guide your understanding". At this juncture, the conversation underwent a sea change – the woman spoke not as subject to the interviewer but as a mentor to the pupil, imparting her knowledge and wisdom on the topic.

In our research, we've sought to give voice to the believers we've interviewed, allowing them to speak for themselves and share their unique perspectives on the world. But as Spivak has pointed out, this is a process fraught with limitations. We've attempted to navigate these limitations by using our own voices as interpreters and translators, bridging the gap between the religious speech of believers and the secular language of academia.

The book is a series of translations, in every sense of the word: from Brazilian Portuguese to English, from individual voices to a more or less cohesive narrative, from the perspectives of Brazilian believers to those of British academia. And it's in this translation process that the concept of "parallax" becomes particularly useful.

Parallax refers to the apparent displacement of an observed object, and it can cause one's view to shift. In the same way, our interdisciplinary research has broadened our understanding by incorporating multiple disciplinary viewpoints. But we recognise that true understanding requires a transdisciplinary approach that challenges and expands our existing frameworks. Through this approach, we hope to grasp better the complexities of the issues we're investigating and shake up the ways we've traditionally thought about them.

Exclusion

To truly comprehend the exclusionary mechanisms at play in society, particularly in urban areas of Brazil, we must first come to terms with the fact that neoliberal cities are built on exclusion. So many bodies and ways of life are pushed to the fringes and peripheries while a dominant centrality holds sway. To fully grasp the nature of exclusion and marginality in urban areas, we must also examine how coloniality is replicated, leading to violent exclusions and silencing of marginalised groups with less power and capital. One example is that the more peripheral an area becomes, the more vulnerable it becomes. As Valencio (2010) illustrates, in urban spaces, disasters disproportionately affect those who are unprepared or unable to overcome them – namely, the poorest. In Brazil, power relations often serve to naturalise these injustices by blaming so-called "natural disasters" for the failure of an inclusive urban model. By designating certain areas as "risk areas" and labelling those who live there as "removable", the responsibility of the state and capital is effectively absolved in debates about urban segregation. The naturalisation of "technical accidents" only serves to justify displacement and removal further. And by blaming victims for their losses and suffering, poverty is criminalised, and the true culprits can evade accountability.

In the era of globalisation, the dynamics of domination and coloniality have been amplified on a global scale, with capital accumulation serving as the driving force behind this expansion. Milton Santos (2001) refers to this as the "imperialist" nature of globalisation. Despite the potential benefits of a more unified world, globalisation imposes itself on the majority of the population in a detrimental way through the "double violence" of the tyranny of money and the tyranny of information. The unequal distribution of information technologies deepens these processes of inequality, and the relationships at the core of neoliberalism also infiltrate the most intimate aspects of our lives, including the body. In the case of Brazil, for example, black people occupy the lowest paying positions, with a hierarchy further compounded by gender, according to Márcia Lima (2001).

In a world where inequality is normalised, as Quijano (2000) notes, it's all too easy to accept and naturalise practices and actions that reinforce this structure. The city, in particular, is built on a foundation of exclusion, using certain central characteristics to distinguish those who inhabit it. As Nascimento (2020) identifies, these characteristics include race, language, gender, land, income, and religion. Of these, religion stands out as a particularly powerful force of exclusion, either through its own oppressive and marginalising tendencies or through the dogmatic structures it enforces.

However, it's important to recognise that religion is also a vast and intricate set of knowledge, encompassing a range of ideas, values, and rules that inform people's understanding of the world. Religion can also be a source of empowerment, giving marginalised groups institutional and symbolic support. According to Martijn Oosterbaan (2022), the strong Pentecostal presence in Brazilian favelas plays both a role in strengthening citizenship (mainly through self-construction) and can also legitimise the authority of violent actors, even portraying criminals as marginalised subjects who, through their evangelical confessions, can claim their citizenship. The author argues that "succeeding to build one's own house and/or a local church with the help of the Lord supports a strong sense of self-value and dignity, which strengthens a sense of equality when confronted with Brazil's socio-economic segregation and stigmatisation".

In our research, we sought to approach participants – the faithful – as gatekeepers and holders of knowledge. By grasping the context in which these individuals are formed and how they perceive themselves as social and historical agents (we use the term "agent" to focus on their agency), we can begin to unpack the complex relationship between religion and the urban landscape.

Brazilianess

Undertaking decolonial research in Brazil is of paramount importance, as the country has a long and troubled history of colonisation and a society that remains deeply ingrained in colonial relations and the notion of race. The legacy of colonisation and the subjugation of millions of enslaved individuals from diverse social, cultural, and religious backgrounds continues to shape class divisions, labour hierarchies, and wealth distribution in Brazil. Christianity has long held hegemony in the country's religious landscape, with little room for other traditions, particularly those of African origin. As per UNDP research from 2019, 56.10% of the population self-identifies as black. The examination of Candomblé and Pentecostalism in Brazil is of particular significance, as Candomblé, a religion of African origin, has a significant presence in the country. The arrival of Candomblé can be traced back to the forced migration of enslaved Africans, who were compelled to practice Christianity but covertly maintained their true beliefs by incorporating Catholic symbols. This has led to the unique syncretism that characterises Candomblé to this day, with some followers embracing both Catholicism and Candomblé simultaneously (Ogunnaike, 2020).

Candomblé offers a deeper understanding of the multifaceted experiences and transformations of black individuals in Brazil, who continue to endure marginalisation and subjugation. In contrast, neo-Pentecostalism, the third wave of Pentecostal expansion in Brazil, is notable for its emphasis on wealth acquisition and incorporating elements that could be considered "magical". These churches compete with the Catholic Church for the conversion of black individuals, yet their message is often at odds with black identity due to the demonisation of African culture and racist undertones in their teachings. In April 2022, for example, evangelical preachers "exorcised" attendees at two African-based terreiros in São Luís (Maranhão), similar to an incident in 2021 that had the support of a state deputy. These events highlight the complexity of the relationship between race, religion, and power in Brazil and the importance of a decolonial approach in understanding these dynamics.

In Brazil, two primary distinctions contribute to ongoing inequality: race and class. Despite having the largest black population outside of Africa, Brazil has a dearth of black representation in positions of power, such as in Congress, universities, and the media. Additionally, there is a high rate of violence directed towards black youth and a strong correlation between poverty and race. As Achille Mbembe (2017) has noted, race relations in Brazil differ from those in countries like the United States. While in the United States, institutionalised racism was established through war and constitutional amendments, in Brazil, the concept of a "pact" has been used to describe the situation. This social pact operates as if slavery and struggles for equality never occurred. It has led to a situation where black individuals pretend they are not discriminated against while white individuals pretend they are not racist. As a result, black people have limited access to education, housing, and representation. Thanks to colonisation and slavery, Brazil remains a racist country, reproducing the treatment that enslaved individuals received with poor people, not just the black community. The importance of understanding and addressing these issues cannot be overstated, as they have deep historical roots and ongoing ramifications for the entire society.

The ongoing coloniality that persists in Brazil – the ways in which bodies are still classified and ordered based on race – is a weighty and intricate issue that contributes to the ongoing inequality and intersectionality of race and class relations in the country. It's a matter that, as researchers and academics, we can't help but feel a sense of urgency about. And yet, as is often the case with these kinds of knotty, systemic issues, untangling it all can feel like a daunting task.

Mbembe's exploration of "becoming black in the world" is particularly enlightening in this regard. He delves into how black men and women have historically been transformed into currency and merchandise by white Europeans. And in more recent times, he examines how social life is increasingly codified by market norms under neoliberalism and how this has led to the abandonment and mistreatment of black people, particularly those in lower classes. And yet, as Haider (2018) points out, the terms "ethnos" and "demos" represent an imagined community of members and affiliations (the former) and a political elaboration that refers to the collective representations, decision-making, and rights of a group (the latter). And when it comes to the debate on identity, it's important to remember that no two are identical or two different.

In Brazil, religion is a fascinating and complex phenomenon shaped by various factors, including race, class, and individual choice. One of the most striking paradoxes is that Pentecostalism, despite its conservative and hierarchical origins, has become a religion of the people, attracting large numbers of women and blacks who are disproportionately represented among the poor and marginalised (Mariano, 2013). At the same time, Candomblé, which has its roots in African spirituality and was once associated with the lower classes, has undergone a transformation, drawing a growing number of white, middle-class followers (Prandi, 2004).

This shift in religious affiliation reflects the broader changes in Brazilian society, where traditional hierarchies and boundaries are breaking down, and people are increasingly free to choose their own path in life. This new religious landscape is also shaped by the diversity of the population, with different cultures and perspectives coming together to create new narratives about the world. And as more and more people turn away from traditional Christianity, we are seeing a growing interest in other spiritual traditions and practices, from Buddhism and Hinduism to indigenous and Afro-Brazilian religions. Overall, religion in Brazil is a dynamic and ever-evolving field shaped by the complex interplay of social, economic, and cultural factors. And as we continue to explore this fascinating landscape, we must be mindful of the many different perspectives and experiences that shape it, and the ways in which it reflects the broader changes taking place in our world.

Process

Given the context outlined above, we set out to conduct a series of interviews that would serve as the foundation for our book. We approached these interviews with humility, recognising the complexity and nuance of the religious worldviews we sought to understand. Our research was divided into four stages, each designed to collect empirical data through various methods.

The first stage of our research occurred in the early months of 2019 when we travelled to the outskirts of Belo Horizonte and conducted in-person interviews with residents of the Eliana Silva and Carolina de Jesus neighbourhoods and students from the Escola Estadual Maria Carolina Campos. Field visits were also crucial to this stage, as they allowed us to understand better the reality of our interviewees' lives.

The second stage of our research was conducted remotely during the first half of 2020, as the COVID-19 pandemic made in-person interviews impossible. We used this stage to test remote interview methods and to ask questions about the spatial practices of religion in the context of isolation and social distancing. While these interviews were initially seen as test cases, they ultimately proved to be a valuable source of information, helping us to better understand the diverse religious practices in Belo Horizonte.

The third phase of our research was also conducted remotely, as the world was still reeling from the pandemic's grip. We employed a "snowball" approach, starting with individuals we had already encountered during our fieldwork and then branching out through their networks to find more participants. The digital realm had become the new frontier, and the virtual connections we forged gave us access to a wider range of perspectives and experiences. As we delved deeper into the data, we began to see the intricate web of connections that linked our interviewees and the ways in which their religious practices were shaped by the ever-evolving landscape of the 21st century.

In the fourth stage of our research, we delved deeper into the workings of Candomblé and Neo-Pentecostalism, conducting in-depth interviews with their followers. We also created visual representations, mapping out the experiences of our interviewees through drawings. Recent fieldwork in 2022 saw us traversing the streets of Belo Horizonte, specifically focusing on the Concórdia neighbourhood, a known hub for Afro-Brazilian religions in the city.

Challenges

During the course of our research, we encountered several challenges. One of which was the language barrier. Brazilian Portuguese and English are vastly different languages, and Candomblé incorporates a significant amount of Yorubá, a language none of this book's researchers could speak. Candomblé is not based on written texts. This means that there is a fundamental difference between Pentecostals, who rely heavily on written sacred texts, and Candomblé believers, who rely more on rituals and oral traditions to practice their religion. Additionally, while there are "non-practising" Christians, there are no "non-practising" Candomblé believers. The interviewees stated that it is quite impossible to practice Candomblé at a distance. They claim they need the exchange with others due to the fact that there's no written tradition and the importance of the temple to the connection with the faith of others. We have learned that Candomblé believers emphasise maintaining sacred spaces and holding meetings, while (Neo)Pentecostals may participate in online or television masses. These observations partly contradict the established opinion about Pentecostals experiencing and practising their religion through their bodies. However, we would like to clarify that there is no clear-cut; followers of Candomblé also read and practice alone, while Pentecostals crave meeting in person.

Another challenge we faced was the transdisciplinary nature of our research. Our group represents a range of disciplines, including geography, sociology, urbanism, architecture, and international relations. This pushes us to consider our work as interdisciplinary. However, as mentioned above, we went beyond academic disciplines and fully engaged with non-academic worldviews and speech. However, it can also be challenging to navigate the diverse backgrounds and methods of analysis represented by our team, which includes researchers at various levels of education, from master students to doctors, and participants with often only a basic level of education. By presenting the narratives and perspectives of our informants as valid knowledge, we aim not only to showcase what they have taught us but also to highlight the fact that each religion constitutes a system of knowledge that shapes our understanding, use, and production of the world.

As we delved deeper into our research, we encountered the complexity of the groups we studied. As discussed previously, these groups were not homogenous, and the diversity of perspectives and experiences presented its own obstacles. For example, within the Pentecostal churches we studied, we found a divide between "inward-looking" and "outward-looking" congregations (we will discuss this issue further in the following chapter), with the former being small and independent, while the latter were often larger and part of a broader network of churches. We recognised that this diversity of perspectives and experiences enriches our understanding of the subject but also makes it more difficult to navigate.

What we hope sets our work apart is a non-dialectical approach, attempting to engage with some contradictory perspectives in a non-exclusionary way, rooted in Ernst Junger's "stereoscopic view" as discussed by Nawratek (2018) in his previous book. By adopting an emancipatory and decolonial approach, we hope to challenge dominant narratives and give voice to marginalised perspectives, providing a more nuanced and holistic understanding of the complexities of religion in the post-secular world.

Final considerations

Our research sought to bring to light the perspectives of those marginalised within Anglo-Saxon and Western academic circles. As (mainly) outsiders to this discourse, with only one member of our team affiliated with a British university and none of us claiming a Western identity, we were acutely aware of the power dynamics at play during our conversations with predominantly Black Brazilians. However, it soon became clear that our privilege as white, educated individuals still positioned us in a distinct place of power. This dynamic was only reversed when we allowed believers to educate us about their beliefs. We have been following David Robertson's framework and recognised our limited understanding of their spiritual practices as equal to other forms of knowledge, such as scientific knowledge or personal experience. By doing so, we could level the playing field and conduct, we hope, a truly co-produced, non-exploitative study. The methods and theoretical framework we employed played a crucial role in shaping this project's outcome. We all underwent a personal transformation through our interactions with Candomble, Congado, and Pentecostal practitioners. Though we did not convert to any of the studied religions, we came to deeply respect the validity of various religious worldviews.

Notes

1 Eduardo Galeano was born in Montevideo, in 1940. With the military regime in Uruguay, he was persecuted for his book *As Veias Abertas da América Latina*, where he analyses the history of Latin America from colonialism to the 20th century. All of his work is important for understanding the history of colonisation and colonialism on the continent.

2 IURD (Portuguese): Igreja Universal do Reino de Deus, the church is known as English-speaking countries as The Universal Church of the Kingdom of God (UCKG).

References

(n.d.). *Apesar do aumento de pessoas negras nas universidades, cenário ainda é de iniquidade.* Gife.org.br. Retrieved January 29, 2023, from https://gife.org.br/apesar-do-aumento-de-pessoas-negras-nas-universidades-cenario-ainda-e-de-desigualdade/

Burley, M. (2017). "The happy side of Babel": Radical plurality, narrative fiction and the philosophy of religion. *Method & Theory in the Study of Religion, 29*(2), 101–132.

Durose, C., Perry, B., & Richardson, L. (2021). Co-producing research with users and communities. In E. Loeffler, & T. Bovaird (Eds.), *The Palgrave handbook of co-production of public services and outcomes* (pp. 669–691). Cham: Palgrave Macmillan.

Galeano, E., & Borges, J. (1993). *Las palabras andantes.* Spain: Ediciones del Chanchito.

Haider, A. (2018). *Mistaken identity: Race and class in the age of Trump.* Brooklyn, NY: Verso Books.

Harloe, M., & Perry, B. (2009). Rethinking or hollowing out the university?: External engagement and internal transformation in the knowledge economy. *Higher Education Management and Policy, 17*(2), 29–41.

Lima, M. (2001). Serviço de branco, serviço de preto: um estudo sobre cor e trabalho no Brasil urbano. Rio de Janeiro: Programa de Pos-Graduacao em Sociologia e Antropologia, IFCS-UFRJ, Tese de Doutorado.

Mariano, R. (2013). Mudanças no campo religioso brasileiro no Censo 2010. *Debates do NER,* 24, 119–137.

Mbembe, A. (2017). *Critique of black reason.* Durham, NC: Duke University Press.

Nascimento, D. M. (2020) *O sistema de exclusão na cidade neoliberal brasileira.* Marília, SP: Lutas Anticapital

(n.d.). *Neopentecostais incitam racismo religioso no Maranhão.* Amazonia Real. Retrieved January 29, 2023, from https://amazoniareal.com.br/racismo-religioso-maranhao/

Nawratek, K. (2018). *Total urban mobilisation: Ernst Jünger and the post-capitalist city.* Berlin: Springer.

Ogunnaike, A. (2020). What's really behind the mask: A reexamination of syncretism in Brazilian Candomblé. *Journal of Africana Religions, 8*(1), 146–171.

Oosterbaan, M. (2022). Rights and stones: Pentecostal autoconstruction and citizenship in Rio de Janeiro. *Space and Culture, 26*(2), 253–267.

Perry, B., & May, T. (2010). Urban knowledge exchange: Devilish dichotomies and active intermediation. *International Journal of Knowledge-Based Development, 1*(1–2), 6–24.

Prandi, R. (2004). O Brasil com axé: candomblé e umbanda no mercado religioso. *Estudos avançados, 18,* 223–238.

Quijano, A. (2000). Coloniality of power and eurocentrism in Latin America. *International Sociology, 15*(2), 215–232.

Santos, M. (2001). Por uma outra globalização: do pensamento único à consciência universal. 6ª Ed. Rio de janeiro: Editora Record

Spivak, G. C. (2015). Can the subaltern speak?. In *Colonial discourse and post-colonial theory,* eds. P. Williams and L. Chrisman (New York: Columbia University Press, 1992), pp. 66–111.

Valencio, N. (2010). Desastres, ordem social e planejamento em defesa civil: o contexto brasileiro. *Saúde e Sociedade, 19,* 748–762.

3 Conquer or hide

The first thought that came to me is that there are basically two spaces. The first, the space that is for communion, which is the space for you to be with other people and for you to talk, worship or, I don't know, eat together. The space where you are with people who share the same faith. The other space I think is just you closed in a room. It could be a bedroom or a living room, but it's a space that I really need, because it's where I want to cry, I want to talk to myself, I want to fight... My space with God and I don't want anyone to come in.

(White female student, Belo Horizonte, Brazil, interviewed by Authors in March 2021)

Impressions

Every morning, a woman walks past me, headphones on and singing gospel songs at the top of her lungs. She sings with her eyes closed, seemingly unaware of me or the city around us, but she sings to be heard. She's doing her part to spread the word of God on her way to work.

During our first visits to the Candomblé terreiros, we were welcomed into a room and given a brief overview of the group's traditions. We were invited to return and learn more about the events and religious services they offered, and, if we were interested, to be initiated into the religion through a series of activities and learning.

As we talked to architecture students, they gradually revealed that they attended evangelical churches and participated in prayer and evangelisation groups but chose not to show this at university. We also met students who had adopted religions of African origin and were open about this at university but were hesitant to reveal this to their families with Christian traditions.

These and other experiences sparked our curiosity about the potential similarities and differences in these behaviours and how they impact people's perceptions, actions, and relationships in the urban space. In this chapter, we explore how Pentecostalism and Candomblé operate within the larger context of Brazil's religious landscape, using two dimensions: "exteriorized-focused" or "outward looking", which seeks to establish and expand the influence of religion into the secular sphere (urban space, politics, institutions, media) and create religious infrastructure as

DOI: 10.4324/9781003248019-3

external support for individual believers; and "introspective-focused" or "inward-looking", which aims to strengthen individual believers' spiritual power and prepare them for the perceived hostilities of the secular world.[1]

We first identified these two categories during our fieldwork in Belo Horizonte in 2019. At that time, in the early stages of our research, our conversations hinted at a dual perception of the role of faith in the context of the urban occupations we were studying. On the one hand, Pentecostal churches' social and political role was evident in daily life, providing a safe space for believers in a hostile environment. In some interviews, particularly those conducted in peripheral areas of the city, people described how faith, prayer, and a sense of closeness to God enabled them to endure the challenges of their environment.

Chapter 5 delves into the sprawling growth of Pentecostalism, which often thrives via the incorporation of smaller temples into larger, more established churches. While these temples maintain their unique denominational identity, they're invited to partake in weekend events and special services, which many find an attractive offer, a way to feel enmeshed in a grander, more organised community, to propagate the divine word with greater fervour. Yet, not all participants are comfortable with the big-tent gatherings orchestrated by these large churches, for such spectacles – although affirming the sense of belonging to a more formidable entity – frequently veer into the political realm, espousing views that do not align with their personal interpretation of faith.

Our fieldwork also revealed the central role that faith and spiritual growth played in daily life, often described as an individual process of connecting with the Holy. During the pandemic, faith was seen as a way to cope with and deal with daily life's uncertainty and social isolation. As one interviewee stated, faith provided both spiritual support and protection, helping them "not to freak out during the pandemic", while religion offered social support and a sense of community in daily life.

In this chapter, we will examine how the "exteriorized-focused" and "introspective-focused" dimensions shape the habits and actions of believers in the contexts we studied. While we will focus mainly on Pentecostalism and Candomblé, we will also draw comparisons with other religious groups, as the boundaries between them are often blurry, and there is often overlap in traditions and practices, as well as relationships of proximity and conflict between different groups, as we will discuss further in Chapter 4.

Exteriorised-focused and introspective-focused dimensions

Pentecostalism and Candomblé have distinct expansion strategies. While Pentecostalism is a religion of individual conversion that seeks voluntary entry into a congregation of believers, it is also characterised by a focus on the future rather than the inherited past, as in the Catholic tradition. This has influenced the content of Brazilian Pentecostalism's exteriorised and introspective dimensions, leading to a series of ruptures in how this religious group operates. In contrast, the shift from ethnic religion to universal religion in Candomblé, as described by Pierucci

(2011), means that "religions linked to ethnicities propagate while cultural ties are alive, which may last for a few generations, but when these ties are broken for various reasons, one of them being the destruction of the ethnic or family ghetto, religion tends to dilute as well". While this topic was only briefly mentioned in our interviews, it seems that Candomblé practitioners are aware of the potential for reconciling their ethnic heritage, syncretic strategies, cultural dimensions, and resistance within the context of their religion and other religions of African origin.

It is worth noting that the expansion of the third wave of Pentecostalism (Neo-Pentecostalism) has resulted in an increase in the number of believers and denominations, as well as a greater focus on pragmatism (less interest in theology and dogma) and the spiritual battle against other religions. Some leaders within this movement have adopted explicit strategies of religious intolerance, particularly towards religions of African origin (Mariano, 2014). The academic explanation for this "holy war of good against evil" often involves the argument that these religions monopolise (so-called) "magical practices" in the "religious market" (Eleta, 1997; Sanchis, 2007; Sansi, 2007). This type of Pentecostalism is characterised by a competitive and combative approach towards other belief systems (Silva, 2006). In response to a process of secularisation and rationalisation that reached global and local Christian sectors, Pentecostalism emerged as a possibility of valuing the experience of religious revival, an experience lived in the body itself, a characteristic also presents in Umbanda and Candomblé.

However, it's important to note that the term "religious market" was not mentioned in this way in the interviews with believers of Pentecostalism and Candomblé, and none of the interviewees directly mentioned the competition over "magic" practices and rituals. Some interviewees suggest that any reported conflicts, which tend to be highlighted in the media, are exceptions and that when they occur, they are promptly denounced and negatively impact religious groups, including Pentecostals (Spyer, 2020). During fieldwork, we heard reports of tolerance between temples located very close to each other and between worshipers from different groups living in the same family. One interviewee, for example, explained that she came from a Catholic background but had recently started attending a Pentecostal church. She said that the pastor had strongly condemned her daughter's cultural habits (such as funk dancing and going to bars), but she saw no problem with her daughter doing these things as long as she had faith in God, prayed, and continued attending church. Another interviewee's mother, the leader of an African-based religion, respected her son's choice to practice Pentecostalism as long as he respected what she considered sacred within their home. According to her, her son always showed respect, and they had good conversations about religion.

Many of our interviewees suggested that this spiritual tension is only compounded by the rise of social media and the proliferation of conspiracy theories and post-truth manipulations, which have been weaponised by some Pentecostal groups to spread their message of intolerance and demonisation of Afro-Brazilian religions. In fact, many of the leaders of these groups actively seek to manipulate the aesthetics of Afro-Brazilian rituals and present them in a way that associates them with demonic figures of the Christian tradition (Mariano, 2014). This has created a sense of fear and mistrust between evangelicals and practitioners of Afro-Brazilian religions, as well as a defensive attitude on the part of those who attend terreiros. Despite this, some interviewees have tried to bridge the gap, inviting evangelical neighbours to visit the terreiro, though without much success. As one interviewee noted, "it is not everyone that I call to enter this sacred space. The person must start by attending open events, then attend prayers and only later participate in the services".

Religious geography of Pentecostalism and Candomblé

If we would dare to make a slightly simplistic statement, we could say that it is clear that Pentecostalism and Candomblé are operating within an unequal power dynamic, with the former seeking to extend its influence while the latter serves as a haven for those seeking protection. This imbalance can be seen in creating places of worship, media representation, cultural legitimacy, and organisational structures within these religions.

During conversations with representatives of an African-Brazilian religion, we gained insight into the shell-like behaviour that characterises these groups and is also seen in the way how their religious spaces are shaped. According to the interviewees, it's essential to be careful about who we bring into the sacred spaces and to consider whether they "vibrate at the same frequency" as the space, show respect for the place and its ancestors, ask permission to enter and express gratitude upon leaving.

Based on this understanding, we believe it's possible to see Afro-Brazilian religions also as expansionist but using very particular strategies, based mostly on the metaphorical use of space and interactions with a broader Brazilian culture. To fully engage with this topic, it's crucial to keep in mind the duality of spatial analysis proposed by Gil Filho (2012). We have to discuss "religious geography" as an attempt to understand the influence of religion on man's perception of the world and humanity and the "geography of religion" as an attempt to analyse the effects of the multiple relationships of religion with society, culture, and environment.

The geography of religion was discussed in Chapter 5, and the highlighting of two dimensions was summarised in Table 3.1.

To advance in understanding the more transcendental dimension (Droogers, 2003, 2010), it is crucial that we examine the symbolic ways in which religious individuals imbue the world with meaning. The challenge lies in identifying the symbolic manifestation of the transcendent, in discovering the perception that allows us to construct an "interpretive bridge between the transcendent and its materiality" (Gil Filho, 2012, p. 16), thereby creating a new way of reading and interacting with the world. To aid in this endeavour, the following table is organised according to the four categories of religious geography outlined by Gil Filho (2012): (1) the religious landscape (the material world as perceived through sensory instruments), (2) the cultural-symbolic system (the logical symbols of religion), (3) scripture and traditions (the group's epistemological constructions), and (4) religious feeling (the impact of the transcendent character) and put in the context of the six epistemic modes discussed by David G. Robertson (Figure 3.1).

Table 3.1 Exteriorised-focused and introspective-focused dimensions on the Pentecostalism and Camdomblé geography of religion

Geography of religion (see Chapter 5)	Pentecostalism	Candomblé
Temples	What we identify: (i) Small temples throughout the territory and (ii) high-capacity buildings in traffic corridors and central locations. Exteriorised dimension: Monumental temples receive thousands of people during the services, increasing the feeling of belonging to a group. Small temples are strategically articulated with the larger churches, giving great fluidity and effectiveness to the expansion of Pentecostalism. Introspective dimension: The individualisation of faith and proximity to God seems to shift the role of the temple to places of greater isolation and intimacy. The temple is used for fraternisation and religious events	What we identify: (i) Older terreiros protected by public policies and open to visitation and (ii) regular buildings and shared with homes. Exteriorised dimension: The geographical expansion of Candomblé does not depend on the capacity and size of the temples but on an internal dynamic. The visitation of terreiros and their introspection are factors linked to protection against religious persecution. Introspective dimension: The physical space plays a fundamental role in the cult, directly associating the spatial elements with the Orixás
Media and Cultural legitimation	What we identify: (i) Large acquisition of radio and TV stations and publishers, and (ii) language for the mass population. Exteriorised dimension: The acquisition of channels and the mass distribution of content are understood as the main instrument for expanding religion. Introspective dimension: Daily and facilitated contact is an essential channel for approaching God and the possibility of spiritual growth	What we identify: Strong cultural capital with intellectuals and the cultural field (greater diffusion of aesthetics than dogmas). Exteriorised dimension: The dissemination of information and images plays an important role in legitimising and counter-narrative of attempts to systematise religions of African origin – most of this material is oriented more towards the protection of freedom of worship than adherence to religion. Introspective dimension: The symbolic content of this type of communication is apprehended in a more specialised way for those initiated in the religion
Political and institutional arena	What we identify: (i) Growth of political capital and the ultra-conservative agenda with direct participation of financial and media support from the main leaders and (ii) massive investment in new temples, training pastors and incorporating smaller units. Exteriorised dimension: Sensation of protection, belonging to a consolidated group, and developing a network of social relationships. Introspective dimension: Sensation of effectiveness of the word of God to deal with the threats of the secularised world	What we identify: (i) Protection of freedom of worship law and historical heritage programmes (recent fragility and ongoing moment of institutional deconstruction) and (ii) trend of articulation between terreiros (to protect themselves) and articulation with social movements. Exteriorised dimension: Pressure for greater visibility, protection of cultural identity, and institutional organisation. Introspective dimension: Even stronger links between the transcendental dimension and the outside world that reinforce the importance of the rites and the ethnic dimension of religion

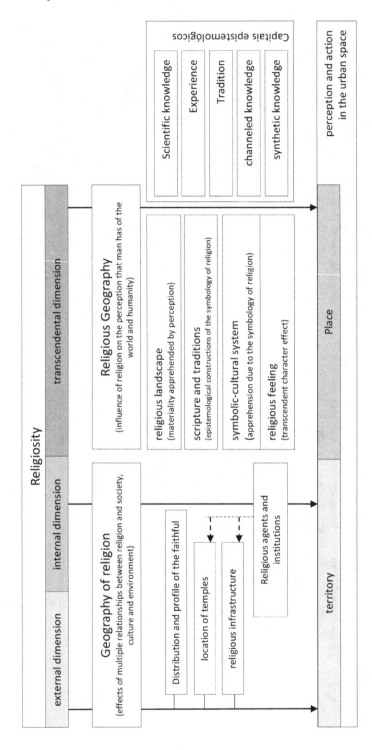

Figure 3.1 The four categories of religious geography.

Source: Design by authors.

In the Pentecostal religious landscape, we see the role played by the sense of belonging to a powerful group whose goal is to become the dominant force in society and increase the trust and visibility of their faith. The size of their temples, the grandeur of their events, and the prominence of celebrities and media outlets reinforce this feeling of being part of something special, chosen by God and blessed by His close presence. At the same time, in practices and rituals, the sense of God's nearness seems more personal and deeply felt by each individual, driven by the emotional appeal and cathartic nature of Pentecostal celebrations.

The material expression of Candomblé, despite the autonomy of each terreiro, is shaped to fit the audience it is intended for. There are invitations for outsiders to come and learn more, but there is also a filter in place to protect against the risk of misrepresentation and decontextualisation of the group's religious practices. In this way, the visibility and identification with cultural and social aspects of this religion are tolerated and sometimes even used to attract new members. But at the same time, the faithful quickly remind outsiders that they are a religion, not simply a form of cultural production or folklore.

For those within the Candomblé community, there is a strong sense of being protected and guided by the Orixás, and a recognition of their role as a source of resistance within conflicts with other religious groups. However, the presence of the sacred also carries its own risks and challenges, such as the need to protect oneself through specific rituals and offerings or the demands of the orixás for protection and respect.

The cultural-symbolic system in Pentecostalism and Candomblé operates similarly, although there are some key differences. Pentecostalism has more effective mechanisms for spreading its symbolic language and has a well-developed infrastructure for this purpose, which allows for creating a more coherent system of interpretation of reality, scriptures, and the perception and actions within the urban space. In contrast, the cultural and symbolic system of Candomblé, due to its strong ethnic character, is closely tied to the traditions and daily life of particular, as mentioned above, autonomous terreiros.

When it comes to scriptures and traditions, the epistemic constructions of the Pentecostal group are based on a more literal and practical interpretation of the scriptures and are heavily reliant on the interpretation of each pastor (who is often not formally trained in theology). In this religious group, there is a strong alignment of cultural and behavioural aspects, with avoidance of contact with non-Pentecostal symbols or behaviours and places deemed profane. There is also a greater openness to adopting certain behaviours, particularly in more loosely affiliated groups. In Candomblé, ancestral rituals and myths are closely intertwined with the group's daily life, and the importance of tradition through ancestry and expected behaviours is a primary reference. Although there is some freedom of interpretation of myths and teaching in this group, as with the Pentecostal group, for Candomblecistas, the knowledge of the elderly tends to take precedence, even when the younger members have greater access to literature and general knowledge. The knowledge in Candomble is hierarchical and achieved by moving upward on a spiritual path led by elders of the terreiro.

Finally, when it comes to religious feelings, both Pentecostalism and Candomblé are characterised by a sense of proximity to the sacred and rituals that involve religious trance and "magic". During interviews, there were numerous reports of these experiences, which are often mentioned in texts on the classification of religious groups. In terms of behaviour in the urban space, this plays a crucial role in the believer's perception of the environment. For example, during a visit to the Concordia neighbourhood, one of the hosts sprayed a liquid on the edges of the public space to welcome us and seek permission and protection from the ancestors to allow us to be in that space. One of the evangelical interviewees mentioned that she avoided streets with many bars because she believed that being in that type of space would distance her from the presence of God.

Epistemic capitals

Our argument is that the "exteriorised" and "introspective" dimensions shape the formation of what we could call a habitus of the faithful. Using Bourdieu's concept, we posit that this habitus is a predisposition, a way of viewing and interacting with the urban sphere that is shaped by, and in turn shapes, an individual's relationship to the objects and actions that make up the city. To put it another way, the faithful forge their own unique perceptions and modes of behaviour, constructing a "personal epistemology" that varies from individual to individual and from one religious group to another.

To understand the formation of this habitus, we can consider, as mentioned before, the work of David G. Robertson, who argues that the legitimation strategies observed in each field of power are shaped by what he calls epistemic capital, which "does not map what you know, but how you know" (Robertson, 2021, p. 28). Robertson identifies six types of dominant epistemic capital: scientific knowledge, tradition, individual experience, institutions, channelled knowledge (knowledge that comes from an external source like gods or spirits), and synthetic knowledge (suggestive narratives created from fragments, he also calls it "assemblage"). By focusing on how knowledge is acquired and legitimised rather than on the content of that knowledge, we can shift the emphasis away from a material analysis of the religious infrastructure (such as the mapping of institutions and concentrations of believers or the demarcation of territories and places in the city) and towards an examination of how religion shapes the constitution of spatiality (a closer approach to religious geography).

In the Concórdia neighbourhood, as described in Chapter 5, the geography of religion gives specificity to the territory through the alignment of the religious infrastructure and the profile of the residents. This specificity is influenced by and influences the perception of followers of African-origin religions, which is reinforced by mechanisms of sacralisation of space related to ancestry, religious symbols, and the presence of religious sentiment in everyday life. This religious geography is shaped by the perception of a landscape filled with symbols on the facades of buildings, tree-lined backyards, and residential architecture that reinforces the traditional ancestral knowledge of some of the residents. In other contexts, this can be seen in the presence of terreiros, which are often more discreet and harder to identify for those who are not initiated into Candomble (or Umbanda). In contrast, visits to areas of urban expansion with precarious urbanisation often reveal visible and active religious geography in the form of small Pentecostal temples of different affiliations, which seem to conform to their own territory based on various functions (such as worship, solidarity networks, and centrality) and amplify their visibility to other religious groups. This geography of religion directly impacts how the resident/faithful perceives and acts within the urban space.

In our conversations in Concórdia, we found that knowledge linked to tradition, always mediated by ancestry and racial issues (a reference to blackness and the African origin of the religion), was the most significant epistemic capital mobilised, highlighting its importance in Afro-Brazilian religions. Both the religious geography of religious symbols (scriptures and traditions) and the way in which these symbols are understood by believers (symbolic-cultural system) were presented to us in a way that was closely tied to the culture of the African people and their resistance during slavery, as well as to the traditional family structure of the group, length of residence, customs, clothing, and cuisine, among other factors. It is also worth noting that other followers of Candomble we interviewed, people who do not live in Concordia and are often white and middle class, operated within similar references.

Returning to the distinction we proposed in the introduction between "exteriorised" and "introspective" movements of religiosity, the tradition observed in this religion represents an "introspective" religiosity (faith as protection and spiritual strengthening) deeply intertwined with the "exteriorised" religiosity manifested through objects and rituals that have been naturalised in the believer's daily life, serving as external protective prostheses. However, the "externalised" aspect is not seen as a priority for expanding the religion but rather as a strategy for asserting and protecting the group's identity. This behaviour is influenced by external factors to which the group is subjected (such as a history of persecution, vulnerability, and prejudice) and internal dimensions of the religion, such as the belief in the spiritual presence of orixás and ancestors in the material world (Prandi, 2001).

In this sense, in religions of African origin, another type of epistemic capital, known as "channelled capital", is more intensely mediated and legitimised by traditional and experiential knowledge than in other religious groups. This characteristic is related to two specificities of the internal dimension of this religion. The first is the autonomy of each terreiro in interpreting the teaching and rituals (Côrrea, 2006; Prandi, 1990), which allows each Mãe and Pai de Santo to be the local authority that defines aspects of interpretation and rules for its practice.

The second is a more complex distinction between good and evil, which expands the space for interpretation and free will among the faithful, particularly when compared to other religious groups. Rabelo (2016) argues that ethics in Candomblé is not simply about assigning value to a ready-made world, but it is inseparable from the process of making beings and the world. This argument rejects the traditional view of Candomblé as an unethical religion, in which the distinction between good and evil is relative to each case (what is good for me in a specific situation may be bad for you and vice versa). Unlike the ethical religions studied by Weber, Candomblé seems closer to the magical and strongly ritualistic religion category, where the production via the ritual practice of exceptional and discontinuous states is more prominent than the cultivation of a continuous type of moral person. In this sense, the author argues that in Candomblé, an ethic is cultivated that turns to situations, seeking in them the key to good conduct: an ethic that is less a system of general principles than an intricate etiquette shaped by an economy of gestures, postures, clothing, and props that mark the position of each one in the terreiro and regulate the relationships between its visitors, not only between humans and entities but also between entities and humans.

We can finally discuss synthetic cognitive capital through the specificity of channelled cognitive capital among candomblecistas. With less institutionalised control and a greater tolerance for individual interpretations, in a world with widespread access to information, the channelled knowledge may be challenged by less experienced practitioners when shared by the elders. A believer who occupies a lower position in the institutional hierarchy may have greater access to information and a wider network for disseminating their interpretations of the religious message. This "personal" dissemination process is not mediated by tradition and may fill the gap between "official" ceremonies and preaching (Prandi, 2001). During interviews and more informal conversations, there were moments when this type of uncertainty emerged, including the potential for generational conflicts, which are ultimately resolved through the belief in tradition and ancestry as the main stabilising force. Another factor that contributes to this stabilisation, particularly in places like Concórdia, is a dense network of neighbourhoods and meetings between groups, which helps to consolidate interpretations and place any dissent within the context of group discussion. However, the autonomy of each terreiro is still respected. In this regard, the custom of periodic parties hosted by other groups, both locally and on a regional scale, also helps to mediate between autonomy and the need for the ability to communicate and collaborate between terreiros. We know this collaboration isn't without tensions, but we are also unaware of any strong conflicts, not animosity, inside the broader Candomble camp.

One aspect we observed during the research that warrants further investigation is how the mechanisms producing particular cognitive capitals differ between followers of Candomble and Pentecostals. It's unclear how exactly channelled epistemic capital is used to question tradition and shape synthetic epistemic capital in these two groups. In the Pentecostal groups, we had contact with, we saw a strong emphasis on the expansion of religion, and relatively less importance was placed on the role of tradition. While expansion is a common feature in Christian religions with a missionary nature, in the Pentecostal group, expansion is presented as a top priority for advancing religion over the secular world, including the conversion of believers of other religions (Mariano, 2017; Prandi, 2004). As for the role of tradition, there is a clear shift towards the future in Pentecostalism that breaks or reinterprets established Christian traditions in other denominations. In this sense, further investigation into the role of channelled and synthetic epistemic capitals in Pentecostalism and how it operates in other religions would complement the research presented in this book.

Turning our discussion towards the analysis of the urban space, the visits and conversations we conducted initially drew our attention to the role of the house as a space for worship, celebration, and residence for followers of Afro-Brazilian religions. In addition to the spatial configuration of Brazilian terreiros, as described in the literature (Côrrea, 2006; Prandi, 1990), in which the limited space reproduces regional conditions on the African continent, the presence of the residence overlaps in a much more complex way with everyday life and religion in the two houses we visited, Dona Neuza and Dona Isabel.

In these two households, family members linger, the kitchen hums with activity, endeavours persist, people and pets traverse the space, and the sounds of daily existence pervade. On the one hand, the atmosphere influences the perception of outsiders, but on the other, it creates a sanctum for the leaders of the community, where they may retreat and nurture their spirituality with greater stability and impulsiveness. To our understanding, this constitutes the pivotal difference between these houses and the terreiros we've witnessed elsewhere, where the domestic routine is less integral to the organisation's infrastructure.

The belief in the presence of ancestors and orixás, beings who perform specific functions and demand certain actions and prayers, is central to the transcendental dimension of this spatiality. This belief allows for fluid territoriality, as the same spiritual presence is carried with individuals to different spaces, blurring the line between sacred and profane. In contrast, the expansion of Pentecostalism is structured through a hierarchy of temples, which can find territories more efficiently than other religious institutions, such as the Catholic church. The interweaving of tradition and experience in the daily lives of those who follow religions of African origin is also a key aspect to consider in analysing epistemic capitals and their relationship with space. Through conversations with Dona Neuza and Dona Isabel, it becomes clear that the life stories of these women and their daily care for the group are closely tied to tradition, with experience converted into wisdom and examples for others.

Finally, the recovery and dissemination of ancestral knowledge through traditional means like herbs, teas, and treatments, as well as art and behaviour, offers a glimpse into the scientific cognitive capital at work in these efforts. We are aware of the shaky place we are here – the "science" we are discussing would not be considered a science by many. However, we would argue that making this leap allows alternative knowledge production based on decolonial intuitions and connected to particular parts of the contemporary academy. Spatially, these efforts and the concepts of place and territory intersect in meaningful ways, particularly regarding the vital study of race and the geography of religion.

Conclusion

Religion serves as a means of grasping the truth and bridging the gap between the tangible world and the symbolic universe. Sacred space (we have doubts if the sacred/profane is the right way to describe space, as we will discuss in Chapter 6) is built out of symbols loosely related to the materiality of space. The practices aiming to preserve or expand the religious territory are rooted in the structure of symbolic space, which permeates and overlaps the material space. The same symbolic structures are present in and constructed by social practices. We believe that our position is – at least partly – rooted in the phenomenological tradition (both Mircea Eliade and Ernst Junger influencing our thinking here); however, phenomenology has already mutated under the pressure of sociological tradition (Weber, Bourdieu, and Durkheim). Still, we believe it benefits a better understanding of the described phenomena if we apply "analysis of the social experience centred on the subjective subject". Social agents qualify and build representation spaces according to collective motivations, such as religious, political, cultural, or economic, in which power is immanent, and the space is a living one with affective connections and a locus of action and experienced situations. Religion adjusts people's attitudes to the "cosmic order" and projects this order into human experience, resulting in dispositions and motivations for social behaviour. According to Bourdieu, religion is responsible for "imprinting a new habitus as a basis for thought, action, and perception, according to the norms of a certain religious representation of the world". We strongly believe it is true, but probably not the whole truth.

Note

1 As we briefly mentioned in Chapter 1, initially, we have been using slightly different dichotomy: political-oriented vs missionary-oriented, however, we believe the typology we suggest in this chapter is better because it is less precise. As the Reader will see, we tend to avoid strong statements, preferring to focus on nuances and tensions while discussing various notions and definitions.

Reference list

de Mello Corrêa, A. (2006). The Candomblé yard: An analysis from the perspective of cultural geography. *Selected Texts from Popular Culture and Art, 3* (1). https://www.e-publicacoes.uerj.br/index.php/tecap/article/view/12620/9798

Droogers, A. (2003). The power dimensions of the Christian community: An anthropological model. *Religion, 33*(3), 263–280.

Droogers, A. (2010). Towards the concerned study of religion: Exploring the double power–play disparity. *Religion, 40*(4), 227–238.

Eleta, P. (1997). The conquest of magic over public space: Discovering the face of popular magic in contemporary society. *Journal of Contemporary Religion, 12*(1), 51–67.

Gil Filho, S. F. (2012). Espacialidades de conformação simbólica em Geografia da Religião: um ensaio epistemológico. *Espaço e Cultura, 32*, 78–90.

Mariano, R. (2014). *Neopentecostais: Sociologia do novo pentecostalismo no Brasil.* Sao Paulo, Brazil: Edições Loyola.

Mariano, A. D. (2017). Resistências ao movimento pentecostal em campanha–mg por um periódico católico. *Revista Discente Ofícios de Clio, 2*(2), 86.

Pierucci, A. F. (2011). Ciências Sociais e Religião – a Religião como ruptura. In F. Teixeira, & R. Menezes(Org.), *As Religiões no Brasil: continuidades e rupturas* (pp. 17–34). Petrópolis-RJ: Vozes.

Prandi, R. (1990). Modernidade com feitiçaria: candomblé e umbanda no Brasil do século XX. *Tempo social, 2,* 49–74.

Prandi, R. (2001). O candomblé e o tempo: concepções de tempo, saber e autoridade da África para as religiões afro-brasileiras. *Revista brasileira de ciências sociais, 16,* 43–58.

Prandi, R. (2004). O Brasil com axé: candomblé e umbanda no mercado religioso. *Estudos avançados, 18,* 223–238.

Rabelo, M. (2016). Considerações sobre a ética no Candomblé. *Revista de Antropologia, 59*(2), 109–130.

Robertson, D. G. (2021). Legitimizing claims of special knowledge: Towards an epistemic turn in religious studies. *Temenos, 57*(1), 17–34.

Sanchis, P. (2007). The Brazilians' religion. *Teoria & Sociedade, 3*(4), 213–246.

Sansi, R. (2007). *Fetishes and monuments: Afro-Brazillian art and culture in the 20th century* (Vol. 6). New York, NY: Berghahn Books.

Silva, V. G. D. (2006). Transes em Trânsito: Continuidades e rupturas entre neopentecostalismo e religiões afro-brasileiras. *As religiões no Brasil: continuidades e rupturas.*

Spyer, J. (2020). *O Povo de Deus.* Hamburg: BOD GmbH DE.

4 Profane space does not exist

Even before I became a pastor, I understood that we cannot mix things; we must respect the limits (…). When I am in the church, I don't mix it with my professional life; I don't take my professional life to the altar to avoid mixing things (…). I cannot cross the social limit, and much less cross the religious boundary (…) "But you are a graduate, you shouldn't do this or that", but inside your formation, you know that there is no wall, there is no barrier, it is more of a mystical thing that people create, that they want to see, at least I treat it like that, I see life as a theologian, as a pastor and as a family father, because I am a family father, right, so I am also a husband and father (…) Look, no, because everyone already knows me. After all, I don't have this business of being one person in the church and another on the street. The same person I am in church, I'm also on the street.[1]

(White man, 49 years old, a pastor in his own church.
This interview was conducted in April 2021)

Sacred (space)

The pastor's words quoted above intrigue us. This recognition of a separation between the spiritual and the secular, yet a simultaneous blurring of those lines within the individual. It begs the question, how can we draw a boundary between the sacred and the profane in the external world if it is not fixed within the very essence of the religious person? A contradiction, perhaps, but one that highlights the fluid nature of the religious self, constantly negotiating and reaching consensus within.

Despite Habermas's interventions (2006, 2008) at the beginning of the 21st century, the Western world still follows a secular perspective in which religious institutions organise the lives of their believers and make public statements on behalf of their members concerning morality, culture, and sometimes politics. God and anything otherworldly is kept in a private sphere for believers. Religious institutions can mediate between their members' personal experiences and beliefs and the secular world of politics, economics, and culture. However, this Western model of a clear division between sacred and secular spheres has been crumbling for many years, if it ever really worked. Even to say that this model is "Western" is not true – the United States has always had religion present in its public life. This sacred/profane distinction would become even more problematic when we go beyond

DOI: 10.4324/9781003248019-4

the West to democratic Asian countries such as Japan or Malaysia. However, that would take us too far from our main topic. Let's focus on Brazil instead.

Pentecostalism is often viewed as an anti-Enlightenment movement that "brings back" the presence of saints, angels, and spirits in people's daily lives. Similarly, Candomblé does not separate the sacred and the profane, as we briefly discussed in the previous chapters and will discuss in Chapter 6. Partly, it is because their deities (Orixás) reside in the world and represent fundamental aspects of nature and society, including technology.

This chapter (and the whole book) is partly indebted to Talal Asad's (2003) complex history of secularity and argues that the world has never been secular for members of the Candomble and Pentecostal churches. For them, the human realm is intertwined with God's/spiritual plane, so the distinction between the sacred and the profane is conceptually inoperative. Therefore, there is a need for a new language to discuss the presence of religious imaginaries in contemporary urban spaces.

The idea of sacred space that has been present in Western discourses is usually seen in contrast to profane or secular space. It is difficult to escape Western duality in thinking. Two (yes, we sometimes fall into this binary model as well) "classic" narratives come from the phenomenological tradition (with the focus on the religious practices and objects as such), represented by Mircea Eliade, and the sociological tradition (with a more functionalist approach to religion), represented by Emil Durkheim. More recently, Kim Knott (2008, 2010, 2015) took a more constructivist position, drawing on the work of Henri Lefebvre and Doreen Massey, while della Dora (2018) introduced the term "infrasecular" in an attempt to overcome the sacred-secular dichotomy. While we believe this is a step in the right direction, it is still deeply rooted in the Western, mostly postmodern, intellectual tradition. We would probably prefer to use the term "infrasacred" rather than "infrasecular", shifting the focus on space seen with a constant connection with transcendence, probably locating our position closer to the initial postsecular intuitions.

But first, let's discuss Durkheim's and Eliade's positions a little bit more. Émile Durkheim understood the distinction between sacred and profane as a universal religious practice. The sacred refers to everything from an extramundane divine realm, and the profane concerns the non-divine earthly plane on which we live. According to Durkheim, religion stimulates social integration because its precepts strengthen those who share the same faith. Thus, religion can be understood as the socially shared experience of the sacred through common beliefs, rules, and authorised forms of behaviour. Durkheim proposes an idea of the symbolic as a mental structure independent of empirical reality. He sees religion as mankind's first system of collective representation, originating in the social domain rather than the supernatural domain. It was the first way of understanding the dual character of the world. Systems of representation emerge from social relations and interact with the categories of understanding that are products of the human mind. The categories of sacred and profane thinking serve as a reference to classify the cognisable world (Montero, 2014).

On the other hand, Mircea Eliade approached religion through the theoretical framework of the history of religion, recognising the sacred stands as one of the most substantial similarities between religious traditions throughout history. Different cultures developed their own cosmologies and worldviews to explain reality and how to interact with the supernatural and divine forces surrounding them. Eliade's view is also influenced by Rudolf Otto (2021), who sought to analyse the religious experience of individuals through their relationship, wavering between terror and awe, with what they understand by "God". The uncertainty and lack of knowledge about the sacred impose themselves on the individual and expose their insignificance before the superiority of this transcendental being, this "absolutely other". From the Durkheimian perspective, which seems to be much more communal, religion would assist in this issue by providing a way of understanding and interacting with the unknown. The social cohesion promoted by religion utilises the aggregated mental processing power of collectivity, pushing the limits of human knowledge and development so that the result is always more significant than the simple sum of the parts. The authors of this book are not free from Durkheim's influence on that issue.

According to Eliade, the sacred's first characteristic is its opposition to the profane. Eliade calls the manifestation of the holy in reality a "hierophany" and states that the history of religions in the world is the history of their hierophanies. When the sacred is identified, it provokes a rupture in the "chaotic homogeneity" of space. What is considered sacred, be it a rock, a tree, or anything else, is treated differently and contrasted to everything else. The sacred represents the foundation of reality itself (Eliade, 1959). The believer lives in a cosmos that is permeated by the sacred and seeks to remain under its constant influence, while the non-religious live in a desecrated, profane world. The sacred/profane dichotomy guides the religious experience while influencing religious individuals' perception and production of space. This is how the Western view of religion has been shaped.

As for our position, we would like to reject the sacred-secular dichotomy when discussing Pentecostalism and Candomble in Brazilian. As mentioned in the introduction, we decided to follow David G. Robertson's (2021) discussion on diverse cognitive capitals, focusing on the mechanisms of world-making used by members of Pentecostal churches or believers of Candomble. Profane space does not exist simply because purely sacred space does not exist either. Space is constantly (re) constructed through the interaction between users (human and non-human actors), geometry, and material objects and then is assessed and evaluated in the context of particular worldviews. The meaning is always diverse and plural – people see space as a concoction of different genealogies, functions, and values. Humans use all sorts of references to give space meaning when understanding space. Not all of these references are rational, even if they are not necessarily metaphysical.

Considering that a central theme of this book is the relationship between religious imagery and the perception and construction of space, we believe that the sacred/profane dichotomy is mirrored by public/private distinction. Both pairs are central to Western discourses, and both are rooted in the same vision of the world. Both pairs could be questionable from several positions, for example, when one puts bodily senses and embodied experiences in the centre of the lived experience. We have been hinting about the importance of the body in previous chapters, but this theme will also be discussed further in this and the following chapters.

(Post)reason

The socio-economic and political transformations in 18th-century Europe marked the beginning of the Enlightenment period, which promoted reason as the only legitimate way of understanding and interacting with the world in emerging modern Western societies. The rational individual that emerged during the Enlightenment period was further refined in later stages of modernity to become more autonomous and independent, while society became more atomised. These processes have been (as we are told to believe) pushing religion to lose its leading role and influence in various spheres of life. As we know now, this description of modernity is only partly true.

The aim here is not to criticise reason but rather the idea that reason is an epistemologically superior faculty and that only through it can one understand the world. At the same time, we want to be very careful in questioning reason. The proliferation in recent decades of conservative and fundamentalist religious movements worldwide, and especially the "birth" of post-truth, has changed the axis of the debate. This is why Robertson's response is so important to us. To understand this further, we need to talk briefly about the secularisation theory, as it still represents an important theoretical starting point for debating the relationship between religion and the modern state. According to a common narrative of secularisation, Western civilisation's development occurred linearly, moving from irrationality and supernatural assumptions to certainties and facts guided by science, leaving no place for religion in a secular society organised by rational principles. The outcome of secularisation should be the worldwide privatisation of sacred/religious practices. But secularisation is not a homogenous process of global reach, as was previously thought. Instead, it has varied according to regional, demographic, and socio-economic characteristics, meaning that elements associated with the secularisation process, such as the displacement of religion to the private sphere, acquire different connotations in each place. There are also significant doubts, brilliantly expressed by Jason Storm (2017), concerning the secularisation of modern science. In Brazil, the categories of public and private are intertwined with the concepts of home and street and influence the behaviour of Brazilians in the different environments they pass through. This discussion will be further developed in Chapter 6.

Charismatics

As we mentioned in Chapter 1, the word "Pentecostal" can cause confusion when discussed in a Brazilian context. The term Brazilian members of Pentecostal churches use to describe themselves is the most often "evangelical", but because of the global and especially North American connotations, this term might be even more confusing. Therefore, for most of the book, we will stick to Pentecostal. One of the very few exceptions we will make in this chapter is to discuss briefly the Baptist church, which from a doctrinal point of view is not Pentecostal, but concerning practices and the general perception of believers, often is seen as such.

First, we need a few words of introduction. Pentecostal or Charismatic churches are rooted in the protestant tradition. Protestantism has a more pragmatic approach to the sacred, privileging a process of portability of the religious experience. Wherever religious individuals are, they can maintain their connection with the divine and often uses their own body as a mediating tool. Pentecostalism – especially its third wave – deepens this trend by prioritising its interactions with the sacred around a great sensorial appeal. Their interactions with the holy and their perception of the body as a thermometer of the presence of the Holy Spirit should be approached from a different perspective than the one used to analyse traditional Protestantism.

A characteristic of the Pentecostal churches is their followers' enchanted vision of reality. Such a posture is at odds with the modern secular narrative that attempts to disenchant the world through reason and science. They sustain "(…) a supernatural vision of natural reality" (Passos, 2020, p. 1122) in which practices and rituals of magical-like interventions are carried out to achieve a certain objective. That enchanted approach towards reality coexists with the modern secular order, influencing their day-to-day practices.

It can be argued that Pentecostal churches go against the tendency of major Western institutionalised religions, characterised by rational control, as outlined by Weber in his religious rationalisation perspective, as is the case with Catholicism in the West. Both Pentecostalism and Candomblé – as discussed below – align with Robertson's approach to the distinct types of epistemic capital employed by individuals. Reality is permeated by supernatural elements and entities that are accessed through a series of supernatural, religious practices, traditions, and rituals.

A perception based on the "theology of the power of God" is behind the whole Pentecostal cosmovision. The birth of Pentecostalism is closely connected to the sensitive manifestation of the power of God, which is recognised by believers following the example of the Pentecost event itself. Such an occurrence also represents empowerment, as it is perceived that an individual, by God's determination, manifests the gifts of the Spirit, such as speaking in tongues or being healed. God works in reality either directly or through those He deems worthy of His gifts and power. "Pentecostal theology is of empowerment sensitively verified by the gifts manifested by the faithful" (Passos, 2020, p. 1124).

The early stages of Pentecostal theology led believers to live a sectarian life separate from the secular world, which was seen as profane. However, that stance has changed over time, and now many Pentecostals aim to transform the world through engagement with it, without fully "belonging" to it. The grace received from God is worn by believers as a sign of their worthiness and establishes a direct connection with prosperity (however, not all Pentecostal follow the prosperity gospel), with success becoming a right granted by God, the master of all material wealth (Gondim, 1993; Mariano, 1999; Passos, 2020).

The way that Pentecostals live their lives is also heavily influenced by their religious worldviews. The Dominion Theology, which aims to prepare the earth for the second coming of Christ, is not just about the political involvement of Christians but also about creating a cultural and social environment that aligns with the Bible. This includes producing books, music, soap operas, films, and radio broadcasts. All of these elements, and the various ways they manifest, contribute to a kind of "presence" in the urban landscape that encompasses both material and immaterial references to Pentecostalism.

In the Baptist Church of Lagoinha in Belo Horizonte, home to the successful gospel band Diante do Trono, a unique strategy can be observed, shared among other Protestant churches with charismatic leanings. As described by Nina Rosas (2015), the band engages in more than just musical performances and concerts; prior to their events, the group holds "intercession seminars". These seminars bring together local religious leaders and collaborators to pray and implore, removing negativity and harm from the community. The members of the church believe that once struggling neighbourhoods have been reinvigorated and improved, both socially and economically, as a result of these efforts.[2] This demonstrates religion's integration of its symbols and imagery into society, going beyond physical structures such as temples and churches to include cultural and societal engagement. Also worth considering is the relationship between this mode of faith and prosperity theology, which views material wealth as evidence of divine favour (de Witte, 2018).

With its forward-looking, expansive ethos, Pentecostalism spreads the sacred in terms of geography and the people's collective consciousness. The city becomes a canvas upon which religious imagery and symbolism are meticulously placed and embedded into the fabric of its structures and the experiences and recollections of its residents. Among some Pentecostals, there exists the tradition of ascending hills to participate in spiritual rituals, as described by Pastor Nilton in a past interview:

"The mount is a kind of tradition in the church. People follow the linear history of the Lord Jesus Christ, who worked during the day, healing and resurrecting, doing his pastoral and missionary work. At night he went to the mount to pray and seek spiritual reinforcement. For example, we are going to start a campaign in the church, a healing campaign. I gather my pastors, the pastors with me, and the missionaries, and we go up there. We have that freedom to talk, to give direction to the work that will be done, the spiritual work, and there we open a kind of purpose. We go to the mountain for three weeks as a spiritual strengthening, so it is a purpose, it is more a place of purpose, it is a place for us to be there having the freedom to pray in private since our lives are so busy during the day, at least for us who work regularly. So when the night comes, we take a specific day of the week to be in this place, which is traditionally a place that people notice to be praying, opening purpose, and doing ministry meetings. At least, I see it that way; now, others may see it differently."

The dichotomy of the physical and the ethereal, the profane and the divine, fails to fully grasp the complexities of Pentecostalism and Candomblé. The "lived religion" approach offers a broader lens, examining the interplay of aesthetics, material and intangible cultural elements, and the aftermaths of colonialism. In a world marked by secularisation and modernisation, one might assume that spiritual expression is confined to designated sacred spaces. But the truth is, this is a limited viewpoint that fails to capture the richness and diversity of religious experience.

Body

The body, a vessel of flesh and bone, is a paradoxical gateway to the divine for believers in Pentecostalism and Candomblé. Through their physical bodies, they experience the manifestation of religious teachings, be it through the writhing convulsions of demonic possession or the tongues of fire that mark the gift of speaking in tongues. During a healing campaign at one such church, pastors and their assistants channel the holy energy through their bodies, acting as conduits for divine grace. One interviewee spoke of the transformative power of faith, a mechanism that provides comfort and solace in moments of crisis and turmoil. As Pastor Nilton observed, faith serves as a balm for troubled hearts and minds, offering refuge from the overwhelming barrage of information in our modern world. The tangible nature of religious experience, the bodily sensations, and material manifestations of faith speak to the very essence of lived religion and its role in shaping the world and its inhabitants.

In modernity, reason reigns supreme as the arbiter of truth, a rationalist framework partnered with an industrial capitalist economy that seeks to harness and exploit nature to produce goods. Space, a precious commodity in this context, becomes parcelled and divided based purely on economic concerns, with little room for subjective interpretation or deviation. But can this state of affairs be disrupted? Once set, concepts and categorisations, such as public and private, secular and non-secular, are ever-evolving and shifting, undergoing reinterpretation and revision as modernity advances. However, despite these artificial boundaries and divisions imposed by economic and political considerations, the boundless capacity of the senses and the sacred embodiment of the human form cannot be contained. Through the body, the secular urban landscape fuses with the realm of the religious and spiritual.

African Matrix

Candomblé is connected to the identity of the black population in Brazil. It is also an essential part of their efforts to assert their ethnic, religious, cultural, and political heritage. Two key aspects must be considered: (1) the importance of the teachings transmitted through oral tradition that form the foundation of their beliefs, and (2) how African religions express and manifest their faith through practices, customs, and rituals. (Pereira & Ferreira, 2018). It's worth making a few observations to further this discussion. As mentioned above, our intuition is that the categories of public/private and sacred/profane overlap with each other. The formation of these categories involves various historical, political, social, economic, and religious processes, and the perceived value of space depends on the lens through which it is viewed. Modern, secular Western societies define public space without references to religious elements. Public space is defined by access, openness, and general freedom of expression. There is nothing to stop followers of any religion from taking over such space to manifest their beliefs. The transformation of public space into a sacred space could happen in an instant.

In Candomblé, the boundaries between religion, popular culture, and black heritage are blurred. Terreiros are considered essential to Candomblecistas for various reasons, ranging from their religious aspect to their role as a cultural space of resistance. However, many Pentecostal groups view terreiros as private spaces of worship and religious practices, rejecting their cultural (and obviously religious) values. Candomble's worldview does not respect the boundaries that have traditionally separated different types of space. Offerings, which are an integral part of Candomblé's practices, can be found everywhere, including outside of the altars dedicated to the deities. It is not uncommon to see offerings in Belo Horizonte – they can be found in the streets, public squares, and woods. Sometimes teenagers may take an alcoholic drink or cigars left as an offering, but they do so with the traditional warning, "be careful, or something bad could happen to you".

In a society marked by spiritual warfare, Candomblé (and sometimes Umbanda) finds itself cast as the archetypal evil, one of Brazil's most oppressed and persecuted religious traditions. In recent decades, the growing concern over violence in academic circles has brought to light the extent of its persecution, which, though once carried out by state institutions and the press, now extends to other actors such as Pentecostal churches and, surprisingly, even organisations dedicated to environmental and animal rights.

At the heart of Candomblé lies the practice of votive sacrifice, which involves feeding the ancestors deemed as deities. While not all practitioners of Afro-Brazilian religions engage in this practice, it has become a key point of contention for those who would persecute the religion, as it only takes a few to perform the sacrifice for all members of these religious traditions to be vilified. Yet, it is worth noting that votive sacrifice is not unique to Candomblé but can also be found in major traditional religions like Islam, Judaism, and Christianity, although in some cases, the sacrificial gesture is symbolic (Prandi, 2000).

Although Candomblé and Umbanda have more visibility today than in the first half of the 20th century, they are still marginalised and discriminated against. This is a consequence of significant changes in Brazilian society, particularly evangelicals' demographic, social, and political rise. Different religious traditions occupy public spaces in Brazil with their respective symbols and activities, but some enjoy greater social prestige than others. Candomblé and Umbanda religious groups have been marginalised for various reasons, mainly due to their ethnic connection with African people and the social stigma of slavery. It's worth noting that the presence and acceptance of African Matrix religions vary depending on the region and community in which they are practised. In Bahia, a state with many descendants of enslaved Africans, Candomblé and Umbanda are particularly prominent. The first ethnic groups to arrive in the region were Sudanese and Bantu, and these influences can still be seen in certain Candomblecist groups today.

The terreiro of Candomblé can be seen as a result of de-territorialisation and re-territorialisation – the forced removal of Africans from their homeland and their subsequent efforts to reclaim their cultural practices and identity in Brazil. These practices and customs extend beyond the boundaries of the terreiro and into the broader urban landscape. In this way, the terreiro can be understood as part of a broader concept of Afro-territoriality, which encompasses the historical, cultural, and political heritage of Afro-descendants in Brazil. Candomblé is closely tied to the identity of the black population and serves as a means of re-establishing their cultural and religious heritage in the face of persecution and oppression.

We need to make, however, two important disclaimers. The number of followers of any religion of African Matrix (Candomble, Umbanda, Congado and other smaller groups) is much smaller than the number of self-defined Black people in Brazil. Also, as we mentioned, there is an increasing number of white Candomblecistas. It's important to note, even if it seems obvious, that slavery in Brazil also meant the suppression of the culture and religion of the enslaved people. There has been a long-standing effort to hide or even deny the influence of African people on the formation of Brazilian society. However, it's clear that well-known Brazilian musical rhythms, such as samba and more recent styles like funk and axé, have been influenced by the black culture of the enslaved and the drumbeats of Candomblé. These cultural influences and blurred boundaries between various spaces and spheres contribute to our argument of the obsolescence of the sacred/profane distinction in the Brazilian context.

It becomes apparent when examining the cosmology and worldview of the Candomblecistas that each Orixá possesses distinct characteristics connected to various aspects of the natural world and the urban environment. According to Yoruba myth, the world was created by Olodumaré, or Olorum, the creator god. The earth, known as ilê-ifé, is the umbilical cord of the universe, the birthplace of humanity and the place from which we spread out. Creating ilê-ifé involved the collaboration of Odudua, who was sent by Olorum to establish the order of creation.

The interpretation of the world as Òrun-Àiyé, a meeting place between the world of the living and the world of the ancestors, is another significant aspect of Candomblecistas' beliefs. At first glance, this might seem similar to the Christian concepts of the sacred and the profane, but it does not represent a dichotomy. Rather, the ancestors are seen as an integral part of the world, interacting with and influencing it. This understanding of the world shapes followers' behaviour, who draw on their mythical and philosophical foundations in their interactions with the world (Melo, 2019). For example, certain locations in the city, such as crossroads, are important to followers of Exu, while a connection to nature is essential for those who worship Oxóssi, and a deep connection to water is vital for devotees of Yemanjá.

One practice within Candomblé involves placing branches of Mariuô (palm tree) on the doors and gates of terreiros to prevent malevolent spirits from entering areas where they are not welcome. Both good and bad spiritual entities are believed to be everywhere, with the spiritual realm permeating all things.

On 8 March 2022, the Celebration of Yemanja took place in Belo Horizonte, a traditional festivity celebrated by Candomblé and Umbanda's followers for the past 60 years. The day was marked by processions and religious activities at Pampulha's Lake, where a site is dedicated to Yemanja, the Orixá of the waters. One attendee emphasised the day's importance, stating that it was a chance for followers of Candomble and Umbanda to showcase their costumes, music, drumming, and singing in public to raise awareness and promote tolerance. This approach could respond to the pressure from Pentecostal practices (some of which we refer to above). However, it is important to stress that the strategy combines religious and cultural activities to mobilise, at least partly, the state's institutions to protect Candomble practitioners.

As we will discuss in detail in Chapter 5, Concordia was one of the first black working-class neighbourhoods to be established in Belo Horizonte, dating back to around 1920. It is considered one of the oldest neighbourhoods in the city, and its history is closely tied to that of the black workers who originally settled there. However, the neighbourhood's demographics have changed significantly over time due to real estate speculation and broader socio-economical factors. From a religious perspective, Concordia is home to a rich and diverse array of African-origin religions, but also Pentecostal and spiritualist (Kardecists') temples. In this neighbourhood, one can find Candomblé terreiros, Umbanda houses, and Congado houses that have maintained their religious and cultural traditions for decades. These religious spaces are deeply rooted in historical, linguistic, aesthetic, and cultural traditions that draw on the African heritage of their followers.

Conclusions

Following Assad, we see secularisation as the privatisation or displacement of beliefs rather than the complete disenchantment of the world. It is not a dialectical relationship in which modern rationality completely excludes religion from society, but rather a dynamic interaction in which secular and religious realities coexist and influence each other. The dichotomy between the sacred and the profane gives a false impression of a strict boundary between them. The notions of "sacred" and "profane" seem misleading and do not help us understand how exactly followers of different religions perceive the world. Robertson's proposition to focus on various sources of legitimisation of knowledge rejects the dominance of reason but does not fall into the trap of irrationality. Robertson helps us to understand that any truth we hold dear is a fragile construct, potentially leading to a more pragmatic definition of the truth or application of Badiou's position while focusing on truth procedures. Any phenomenon or event is put into a broader context and analysed by its consequences.

The focus on context and consequences, asking what a particular phenomenon does instead of asking what it is, may look, on the one hand, too complex. On the other hand, it may look like an easy way to avoid difficult questions. However, this approach allows us to be more nuanced and more open in discussing how space is perceived by followers of various religions. It will enable us to pay more attention to performative aspects of faith, such as music, aesthetics, gestures, and the value attributed to bodies. DaMatta's concept of the "house" and "street" as a metaphor for different ways of thinking about Brazilian society highlights the importance of locality in any analysis. As discussed in Chapter 6, the word "rua" used in Brazilian Portuguese does not translate precisely to "street" in English.

Droogers proposes an approach that privileges how the faithful see and interact with the world from internal, external, and transcendental categories. Even if this approach looks more simplistic than that used by Robertson, it forces us to go beyond the slightly Latourian way of equally evaluating various sources of legitimisation of the knowledge and propose a more vertical and scalable approach to analyse any religion-related phenomenon. Even if there is a certain tension between these two perspectives, we are happy, following Jungerian' "stereoscopic view" to embrace it.

What is evident in both Robertson's and Droogers' approaches is the emphasis on the relationships established between individuals and the sociocultural and religious elements they interact with. The body, understood holistically, moves between different spaces with various forces, energies, and spiritual entities. We may say that profane space does not exist, but what is really important is to understand the fragility of any concept and definition. What really matters is the consequences and context of believers' actions.

Notes

1 The Pastor is a micro-entrepreneur in the field of IT and cell phones. He has a degree in theology, a bachelor's degree in Christian leadership within pastoral theology, and has held the diaconate.
2 One member of the research team had an informal conversation with a former member of the band who states that these transformations have occurred in the places where they performed. This is also the perception of those directly involved in these activities: http://mixgospelblog. blogspot.com.br/2011/12/comunidade-crista-logos-pretende. html, accessed 7 February 2023; another example given is in Feira de Santana, in the state of Bahia. See more about this at: https://youtu.be/b-Oa6hIVafg, accessed 7 February 2023.

References

Asad, T. (2003). *Formations of the secular: Christianity, Islam, modernity.* Stanford, CA: Stanford University Press.
De Witte, M. (2018). Pentecostal forms across religious divides: Media, publicity, and the limits of an anthropology of global Pentecostalism. *Religions, 9*(7), 217.
Dora, V. D. (2018). Infrasecular geographies: Making, unmaking and remaking sacred space. *Progress in Human Geography, 42*(1), 44–71.
Eliade, M. (1959). *The sacred and the profane: The nature of religion* (Vol. 81). Boston, MA: Houghton Mifflin Harcourt.
Gondim, R. (1993) *O evangelho da nova era: uma análise e refutação bíblica da chamada Teológica da Prosperidade.* São Paulo: Abba Press.
Habermas, J. (2006). Religion in the public sphere. *European Journal of Philosophy, 14*(1), 1–25.
Habermas, J. (2008). Notes on post-secular society. *New Perspectives Quarterly, 25*(4), 17–29.

Knott, K. (2008). Spatial theory and the study of religion. *Religion Compass*, *2*(6), 1102–1116.

Knott, K. (2010). Religion, space, and place: The spatial turn in research on religion. *Religion and Society*, *1*(1), 29–43.

Knott, K. (2015). *The location of religion: A spatial analysis*. Abingdon: Routledge.

Mariano, R. (1999). *Neopentecostais: sociologia do novo pentecostalismo no Brasil*. São Paulo: Edições Loyola.

Melo, E. C. (2019). Orun-àiyé-o "sagrado vivido" pelos membros do candomblé e a afro-territorialidade: diálogos entorno de um campo de possibilidades. *Revista Relações Sociais*, *2*(3), 0477–0491.

Montero, P. (2014). Religion, ethnicity, and the secular world. *Vibrant: Virtual Brazilian Anthropology*, *11*, 294–326.

Otto, R. (2021). *The idea of the holy: An inquiry into the non-rational factor in the idea of the divine and its relation to the rational*. Eugene, OR: Wipf and Stock Publishers.

Passos, J. D. (2020). Uma teocracia pentecostal? Considerações a partir da conjuntura política atual. *HORIZONTE-Revista de Estudos de Teologia e Ciências da Religião*, *18*, 1109–1109.

Pereira, R., & Ferreira, F. A. (2018). Visibilidade e Representações Sociais do Candomblé pelo Jornal do Brasil e Correio da Manha (1950–1990). *Um estudo de caso sobre o terreiro da Gomeia e seu dirigente (Duque de Caxias/Rj). Historia, Religiões e Religiosidades na America do Sul*, *17*(2), 233–256.

Prandi, R. (2000). African gods in contemporary Brazil: A sociological introduction to Candomblé today. *International Sociology*, *15*(4), 641–663.

Robertson, D. G. (2021). Legitimizing claims of special knowledge: Towards an epistemic turn in religious studies. *Temenos*, *57*(1), 17–34

Rosas, N. (2015). "Dominação" evangélica no Brasil: o caso do grupo musical Diante do Trono. *Contemporânea-Revista de Sociologia da UFSCar*, *5*(1), 235–235.

Storm, J. A. J. (2017). *The myth of disenchantment: Magic, modernity, and the birth of the human sciences*. Chicago, IL: University of Chicago Press.

5 Religious infrastructure

It's a church we have been working on for about eight years. It's not so new, but it is a neighbourhood church, a community church, so it's a small church; it's not a big church; we have no affiliations with nominations, with anything, so we are an independent, neighbourhood church, that cares for families and family conflicts, social conflicts that we have many there, so our work, the local church, is more like this.

(Nilton, white man, 49 years old, a pastor in his own church. This interview was conducted in April 2021)

Religion landscape

In the early decades of the 20th century, Brazil was widely perceived as a predominantly Catholic country, with Catholic churches occupying central areas and playing a critical role in shaping the urban landscape. However, this picture changed dramatically in the latter half of the century with the rapid growth of Pentecostal churches and the re-emergence of Candomble, a religion with African roots. Unlike Catholic churches, Pentecostal churches are often located in areas of urban sprawl, along traffic corridors and in high-traffic areas, while Candomble houses are typically hidden away from busy city areas.

The decline of Catholicism, the growth of Pentecostalism, and the struggle for preserving Candomble houses have sparked a wave of research in religious geography and urban planning. The literature (Alves, Cavenaghi, Barros, & Carvalho, 2017; Arenari, 2015) has primarily focused on two main areas: (i) the demographic distribution of religious followers and its correlation with the urbanisation process, and (ii) the intra-urban distribution of religious institutions and their relationship with the surrounding urban context.

While these approaches offer valuable insights into religion's material and functional effects on the urban landscape, our aim in this chapter is to examine the additional dimension of how religious beliefs and worldviews shape the perceptions and actions of believers. By considering this dimension, we hope to gain a more nuanced understanding of the complex interplay between religion, urban space, and human experience in Belo Horizonte.

DOI: 10.4324/9781003248019-5

Brazil and Minas Gerais demographic distribution

Brazilian census data from the turn of the 21st century provides a fascinating glimpse into the unequal distribution of followers of various religions across the country. A closer examination of the data reveals a stark contrast between the percentage of Catholics and Protestants (which includes Pentecostal churches) in different regions. The map of the percentage of Catholics shows that cities with a high proportion of Catholics are primarily located in the Northeast, Southeast, and Southern regions, with a noticeable decrease in coastal cities, which are home to the main urban centres of these regions. In contrast, the map of cities with the highest percentage of Protestants reveals a coincidence between the main urban centres of the aforementioned regions and the country's central-west and northern regions. These findings highlight the complexity of religious distribution in Brazil and offer valuable insights into the distribution of religious institutions and their impact on the urban landscape. As we delve deeper into the relationship between religion, urban space, and human experience, it is important to keep these patterns of national religious distribution in mind.

The data from the census of 2000 and 2010 reveals a pattern of growth in Pentecostalism, which increased from 10.4% to 13.3% over the course of the decade. This growth was particularly pronounced in the North and Midwest regions of the country, which points to a greater concentration of Pentecostals in less urbanised areas or areas undergoing more recent processes of urbanisation, as opposed to the historically urbanised regions of the Northeast, Southeast, and South. This distribution reflects the close association between the growth of Pentecostalism and the expansion of agribusiness in the North and Midwest regions. This association is influenced by the political and ideological proximity of the Pentecostal groups with the groups of landowners and entrepreneurs linked to agribusiness, who have consolidated a powerful political group with a wide range of agendas, including pressure on relaxing environmental regulation, the weakening of protection of indigenous lands, and the alignment of conservative agendas with the interests of local agents. However, it is important to note that this reading does not simplify the complex expansion of this religious group. Pentecostal growth has been observed in almost all municipalities in Brazil, reflecting a broader cultural, political, and economic transformation shaping the landscape of religion in the country. As we delve deeper into the relationship between religion, urban space, and human experience, it is crucial to understand the multifaceted drivers of Pentecostal growth in Brazil (Alves, Cavenaghi, Barros, & Carvalho, 2017).

The state of Minas Gerais, home to the city of Belo Horizonte, boasts a higher percentage of Catholics than the national average (70.43% against 64.63%). Within Belo Horizonte, the percentage of Catholics is slightly lower (59.87%) compared to the national average for urban areas in Brazil (81.16%) and Minas Gerais (82.28%). This distribution can be attributed to the strong Catholic tradition in the smaller, older cities of Minas Gerais, where the Catholic presence is more significant in the main urban centres. These cities underwent their urbanisation process during a time when the economic, political, and cultural power of the Catholic Church was much stronger and more prominent than it is today. The Catholic Church built and managed an extensive network of institutions, including hospitals, schools, and housing, as well as a large number of temples. The rapid urbanisation of Minas Gerais in the 18th century, driven by the gold mining industry, was shaped by the political and economic power of the Catholic Church. A stroll through the cities of Minas Gerais reveals the impressive territorial importance and wealth of construction of the temples in all of the region's historic towns. This spatial context also contributes to the lower growth of Pentecostalism in small and older cities in other states in the Southeast region, where the main urban centres tend to have higher growth of Pentecostalism than the interior regions.

A closer examination of the data for the Pentecostal population reveals significant differences between the state of Minas Gerais and the national data for the top five Pentecostal denominations. According to the 2010 census, the rate in Minas Gerais is lower than the national rate for the Assembléia de Deus (3.64% compared to 6.46% nationally), Igreja Universal do Reino de Deus (0.04% compared to 0.98% nationally), and Congregação Cristã (1.08% compared to 1.20% nationally). However, the rate is higher than the national rate for Igreja Quadrangular (1.77% compared to 0.95% nationally) and Deus é Amor (0.81% compared to 0.44% nationally). The most likely explanation is that these leading Pentecostal denominations experienced their initial growth primarily in their states of origin before expanding to other regions. However, it is important to note that these data are now several years old, and while updated data will not be available until 2023, there are indications that Pentecostal growth has continued and may even be more significant. For example, research published by the Datafolha Institute in 2020 showed a decrease in the percentage of Catholics in the Southeast region of Brazil, which includes the state of Minas Gerais, from 60.76% to 45%, while the percentage of Protestants increased from 25.9% to 32%. Unfortunately, we do not have specific data for the city of Belo Horizonte, but the mapping of religious temples presented later in this chapter does provide some indication of significant growth within the Protestant group in the city. As we delve deeper into the relationship between religion, urban space, and human experience, it is important to keep in mind the diversity and complexity of the Pentecostal population in Minas Gerais and beyond.

The data from the 2010 census shows that the percentage of religions of African origins, such as Candomblé and Umbanda, is lower in Minas Gerais compared to the national proportion. According to the census, only 0.21% of the national population identified as Umbanda, and 0.09% identified as Candomblé, with even lower percentages in Minas Gerais (0.07% for Umbanda and 0.02% for Candomblé). This disparity can be explained by the greater concentration of religions of African origin in the Northeast region and the phenomenon of religious syncretism. As previously noted, religious syncretism, or the blending of different religions, has been a pervasive feature of the religious landscape in Brazil. In regions like Minas Gerais, where the Catholic Church held a great deal of power during the period of slavery in the 19th century, enslaved Black people developed strategies to maintain their own religions while also practising Christianity. Even though the percentage of people who identify as Umbanda or Candomblé may be small, the syncretic blend of these religions with Catholicism can give them greater visibility and popularity, often as a secondary religion. One example is the presence of Congado groups, which blend African and Catholic cults in syncretic ways and have a strong presence in cities throughout Minas Gerais. These groups self-describe as part of the Catholic Church and demonstrate the complex interplay between African-originated religions, Catholicism, and religious syncretism in Minas Gerais and beyond.

The data reveals a clear correlation between demographic information and the urban context regarding Pentecostal growth, which tends to occur primarily in peripheral areas around major urban centres (Arenari, 2015; Becceneri, de Farias, & Chiroma, 2019). This supports the commonly held association between the general profile of Pentecostal believers and unequal urbanisation in Brazil. In this chapter, we will examine how this correlation between religion and social precariousness can be viewed through a materialistic and functional lens and how Pentecostal growth can benefit from these conditions. The distribution of demographics for religions of African origin, such as Candomblé and Umbanda, correlates with the black and low-income population. However, these religions are often viewed through a different materialistic lens, characterised as cultural and social resistance. This profile is largely due to these religions' cultural and identity significance and their role in discussions around racism in Brazil. For instance, many of the groups we encountered regularly host cultural events and are connected to social movements and agendas related to anti-racism, feminism, and inclusivity. However, in our conversations with representatives of these religions, we noticed a desire to be recognised as a religion beyond a materialistic analysis of their history and interactions with communities. Despite acknowledging their cultural and political importance, they consider it vital to be acknowledged as a religion, separate from any materialistic (social, political, economic, and cultural) analysis. This highlights the complex intersection of religion, urbanisation, and identity in Brazil and the importance of considering these factors in our analysis.

Studies have shown that the Pentecostal Church can play a mediating role in addressing urban crises by promoting non-violent behaviours (especially related to domestic violence and violence against women), providing social assistance, and creating networks of solidarity and problem-solving (Oliveira, 2012; Rocha, 2019). Juliano Spyer (2020) also highlights how conversion to Protestantism, particularly among black and brown populations, can lead to upward social mobility and greater representation in traditionally white, Catholic elite spaces in Brazil. The evangelical environment promotes personal discipline and resilience in its followers, encourages a culture of entrepreneurship, strengthens mutual aid networks, and supports investment in professional development. Our research also revealed a tendency among Pentecostal believers to prioritise their connections with their religious group and avoid interactions or spaces that might distort them from their faith. This behaviour creates a Pentecostal belt", where evangelical believers tend to distance themselves from other religious groups and seek spiritual growth only through interactions with their own religious community (Mafra, 2011). This often involves studying the same texts, listening to the same sermons, following the same leaders, only listening to worship music, and attending events and places exclusively with others from their own group. However, it is important to note that there is significant internal ecumenical diversity and fluid loyalty among Protestants, with many individuals attending different temples or switching between churches. This diversity within the Protestant (and especially Pentecostal) community highlights the complexity of religious beliefs and practices and the fluidity of religious identities in the urban landscape of Brazil.

Belo Horizonte temples distribution

In the sprawling urban landscape of Belo Horizonte, the distribution of religious groups and their respective places of worship is a complex and dynamic phenomenon. Our mapping of religious temples in the Metropolitan Region of Belo Horizonte (RMBH) and the city itself has revealed many new, mainly Pentecostal, temples located along peripheral urbanisation vectors. This indicates a growth in the number of worshippers that may not be fully reflected in the numbers of the 2010 Census. Our ongoing classification and confirmation of each temple location have thus far revealed a diverse religious landscape, with 847 Pentecostal temples, 447 other Protestant temples, 255 Catholic temples, 186 Candomblé-Umbanda temples, 30 Kardecist temples, and 25 temples belonging to other religions. This information offers important insights into the interplay between religion and urban dynamics and the role of religious institutions as agents of change in the city (Figure 5.1).

Belo Horizonte's burgeoning population of 2.7 million inhabitants is crucial to acknowledge the dynamic of urban expansion that primarily takes place beyond the city's boundaries proper and into its peripheries and neighbouring towns. Here, the fervour of Pentecostalism is felt with greater force. To comprehend the maps we have produced, it is important to consider two key indicators. The map (Figure 5.2) on the left depicts the evolution of the urban area over time, reflecting the city's origin as a planned entity at the close of the 19th century, initiated in the current central area and gradually extending along the East/West axis of the industrial-railway corridor, before incorporating new zones of urbanisation to the North and South. The right map, on the other hand, presents the index of urban quality of life-based on data from the year 2000. It is clear from this map that the central area boasts a superior urban infrastructure and availability of public facilities. The Pampulha neighbourhoods to the North and Barreiro to the South, both of which are centralities with a favourable quality of life index, are situated in pericentral locations.

When considering the locations of Catholic temples, a clear pattern emerges: they tend to be situated in older, well-established neighbourhoods. It is worth noting that the map does not account for other important components of Catholic infrastructure, such as administrative buildings and institutions like hospitals and schools. The mapping of these religious structures reveals a uniformity in their placement, which can be attributed to the tradition of organising territories through parishes, as articulated by Gil Filho (2006, 2012). These parishes serve as the tangible manifestation of the Church's evangelising efforts. Even in more densely

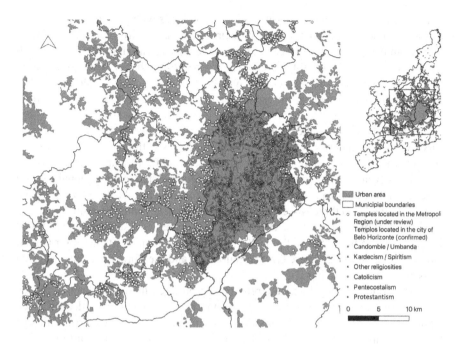

Figure 5.1 Mapping of addresses of temples and churches in the Metropolitan Region and temples already confirmed within the municipality of Belo Horizonte.

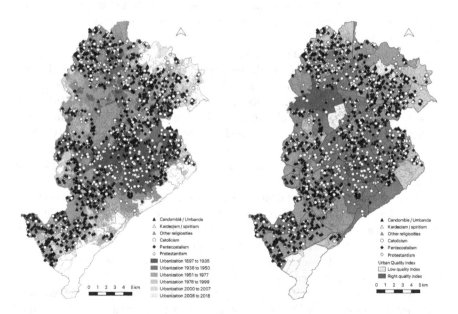

Figure 5.2 Urbanisation process and Urban Quality of Life Index in Belo Horizonte-MG.

Source: Elaborated by the authors.

populated areas, the temples are often situated on lands with greater visibility, size, and value, occupying a central role in their immediate environs. The main deviations from this pattern can be seen in the newest zones of urbanisation in the North and South of the municipality, which also happen to be the areas with the lowest quality of urban life. This discrepancy may be due to Catholicism's slower pace of temple building compared to the rapid expansion of Evangelical religions, particularly the Pentecostals (Figure 5.3).

In the study of Evangelical temples, as depicted in Figure 5.4, it became necessary to distinguish between the various denominations within this broad classification. On the one hand, there were the non-Pentecostal Protestants, which included Baptists, Adventists, Lutherans, Presbyterians, Methodists, Congregationalists, and others. On the other hand, there were the Pentecostal Evangelicals, which could be further divided into three waves of expansion: Classical Pentecostalism (including Christian Congregation and Assembly of God churches), the second wave (marked by growth from the 1950s to the 1980s and the establishment of Brazil for Christ, God is Love, and Foursquare Gospel churches), and the third wave, also referred to as Neo-Pentecostalism.

Machado (1996) highlights the relationship between these temples and the urban environment, noting that Pentecostal churches tend to employ a decentralised strategy regarding their regional distribution. Unlike the more fixed territoriality of Catholic churches, this mobility of temple spaces results in a more informal and ephemeral territorialisation within urban areas, a phenomenon evident in our mapping of the temples and our fieldwork observations. Furthermore, the architectural typologies of the identified Pentecostal temples often reflected this fluidity, with signs of recent renovations, makeshift construction solutions, and the occasional utilisation of existing buildings.

According to Machado, a close relationship exists between the territorial infrastructure of Pentecostalism and its theological and institutional framework. The author outlines four primary components of this infrastructure: (i) supralocal bodies with regional, national, or even global reach that oversee the churches; (ii) large-scale headquarters or mother churches; (iii) branch churches, typically situated in rented spaces such as rooms or halls; and (iv) preaching locations or spots, which are often temporary in nature. This internal structure enables the temples to have a nucleated spatial structure, a primary means of expansion employed by different denominations within the faith.

Our conversations with residents of peripheral areas of Belo Horizonte revealed a fascinating dynamic in expressing their religious beliefs. A single individual might attend a small local temple on a daily basis while also participating in larger, more regional temples on weekends and special occasions.

The distribution of Evangelical temples on the map is less uniform than that of Catholic churches, with areas of lower temple density in the central regions of Belo Horizonte (where urbanisation is more established) and fewer temples in areas with a higher quality of urban life. However, there is a trend towards the location of regional and more prominent Evangelical temples in these areas, playing a crucial role in the expression of religion and capable of accommodating larger numbers

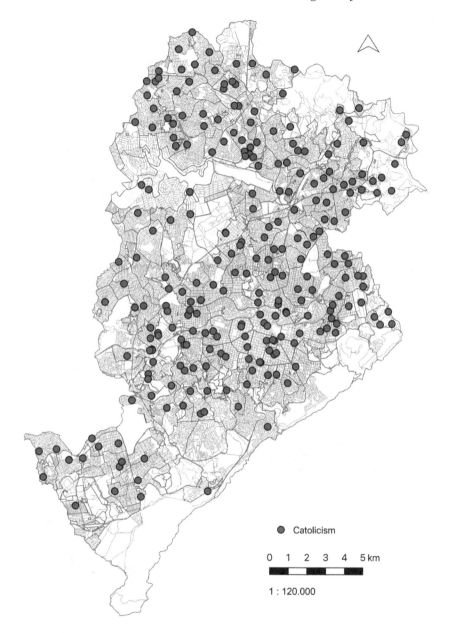

Figure 5.3 Catholic temples in Belo Horizonte-MG.

Source: Elaborated by the authors.

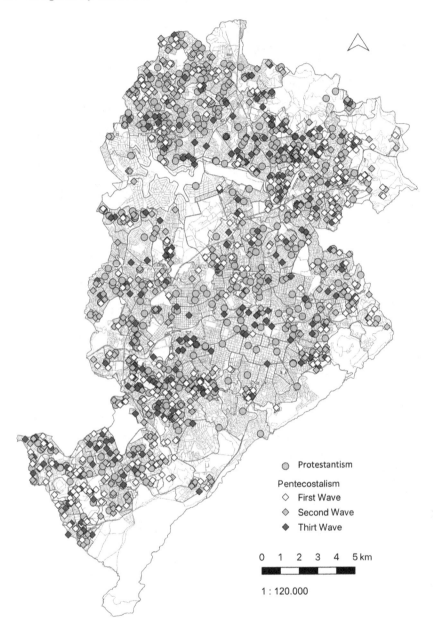

Figure 5.4 Evangelical and Pentecostal temples in Belo Horizonte-MG.
Source: Elaborated by the authors.

of believers. These temples are often situated near major traffic corridors. Four urban areas showed the highest concentration of temples: the Northwest, Northeast, Southwest, and South of the municipality, where the newest zones of urbanisation are located. As previously noted, mapping the metropolitan region would likely reveal even more Pentecostal temples in neighbouring municipalities, especially those undergoing recent peripheral urban expansion (Figure 5.4).

In mapping the temples of religions of African origin, as well as Kardecist (Spiritist) temples and other non-Christian religious groups, it becomes evident that there are significant locational differences between these groups and between them and Pentecostal churches. The efforts of the Belo Horizonte City Hall[1] in mapping the terreiros of African-based religions were invaluable in bringing these spaces to the forefront. However, our fieldwork revealed that it can be challenging to accurately confirm the locations of these temples, as many of the plots of land identified by the City Hall as Candomble terreiros appeared to be for sale or empty.

The territoriality of African-based religions is shaped by the enslaved people brought from Africa to Brazil who established territorial support for the myths and cults practised by their ancestors. While these myths were once associated with broad areas in their regions of origin, they were confined to more restricted spaces in urban Brazil. According to Moura (2019), terreiros serve as the material space or settlement for sacred energy (Axé). Unlike the temples of other religions, terreiros often lack distinctive external features or identification, blending in with the surrounding residential buildings. However, internally, terreiros have a hierarchical structure, usually organised into three classes of myths: the orixás of the frontier/ gate, the orixás of the house/barracão, and the orixás of the bottom/bush. In this organisation, most religious services, visits, and offerings occur in the house/shed.

The micro-insertion of Candomblé, in addition to the various factors that shape its territorialisation in Brazil, is also influenced by the repression it has faced throughout history. This repression has resulted in a more defensive posture for the faith and a more guarded relationship between the buildings and their immediate surroundings. Despite the guarantee of freedom of worship by Brazilian legislation, laws in the 19th and 20th centuries prohibited the identification of temples of African origin, leading to their stigmatisation. With the rise of Pentecostalism and the resurgence of hostilities towards spaces associated with Candomblé and Umbanda, this defensive posture has had to be reasserted in response to attacks on terreiros across Brazil.

On the other hand, various initiatives to map the infrastructure of Candomblé and Umbanda, linked primarily to cultural and social protection policies, have contributed to the visibility and protection of leading centres. However, our visits to these locations left us with the impression that most of them prefer to remain enclosed and without external identification. When visits are allowed, they are always guided and mediated by predetermined itineraries, clearly demarcating the space for visitors from the space reserved for those seeking religious services or those already initiated in the cult. It is noteworthy that the distribution of terreiros in the city is also influenced by the land structure, with neighbourhoods with a tradition of Afro-Brazilian religions often hosting a series of religious street parties alongside

Catholic festivals. As we will see in our analysis of a specific neighbourhood, the location of terreiros, while maintaining independence and autonomy from one another, plays a crucial role in their visibility and preservation.

Concerning Kardecist (Spiritist) temples – they tend to be located in central areas or neighbourhoods, with a high level of urbanisation, reflecting the group's middle and upper-middle-class profile and its establishment in the early stages of urbanisation in the city. However, it is important to note that some temples of African origin are also registered as "spiritual centres", such as the Kardecists. This may lead to inaccuracies and ambiguity in the mapping.

The few temples belonging to other religious groups typically serve as regional headquarters for the religion or provide religious services in small buildings. Their locations are chosen for easy access for worshippers (Figure 5.5).

Temples and the urban context

When examining the typology of temples and their micro-urban insertion, only a few authors (Arenari, 2015; Mafra, 2011) have considered the spatiality of religion beyond demographic dimensions and its correlation with urbanisation processes, particularly in terms of the architectural and urban elements that shape the insertion of temples. This study aims to bridge this gap and gain a deeper understanding of the urban space resulting from the micro-insertion of different temples, focusing particularly on Pentecostalism and religions of African origin and how these groups differ from the Catholic tradition.

By focusing on this scale, we are able to delve into how religious groups establish direct relationships with urban daily life and mediate between the faithful, the city, and religious groups. During our conversations with believers and field visits, we observed that religion directly impacts how individuals perceive and interact with urban spaces. In addition to shaping territories and borders based on the city's distribution of temples and worshippers, religion creates places based on diverse worldviews and ways of experiencing the urban environment. This is the dimension we will explore in more detail in this section, describing the insertion of temples at the local scale and analysing a neighbourhood where the religious territoriality of African origin and its borders with Pentecostalism offers valuable insight into this approach to the topic.

In terms of their insertion into Brazilian cities, Catholic temples are characterised by their central role in urbanisation, occupying symbolic and economic power positions. This is manifested through their high-visibility construction, often accompanied by complementary and institutional structures such as parish houses, schools, hospitals, daycare centres, and community halls. These spaces serve as the hub for religious events and a significant portion of the daily life of the surrounding urban area. This configuration is more pronounced in smaller and older cities but tends to weaken in large urban centres. In Belo Horizonte, the presence of Catholic churches is particularly prominent in the planned areas of the city. The main temples are often detached from the urban landscape and surrounded by supportive structures in central neighbourhoods. In larger temples, there is usually a square or

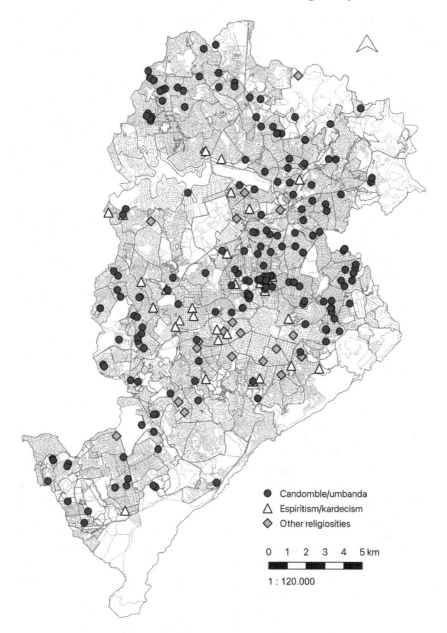

Figure 5.5 Candomblé, Umbanda, Kardecist and other religious temples in Belo Horizonte-MG.

Source: Elaborated by the authors.

setback from the street, while smaller Catholic temples often have entrances facing the road, a typology shared by many traditional Protestant churches, Jewish temples, and Islamic mosques in those neighbourhoods. The area surrounding major churches often serves as a mediator between the temple and the city, hosting church activities or providing a space for socialisation for the faithful before and after worship. In some cases, this area also functions as a socialising space or square for residents and regulars, with little interaction between these two uses. In the case of temples without this type of space, meaning those with entrances facing directly onto the street, there is usually a small atrium serving as a boundary, with several buildings also featuring a second, more internal space for socialisation before and after services.

The insertion of Pentecostal churches in Belo Horizonte differs significantly from that of Catholic churches. Pentecostal churches tend to have at least three different spatial typologies. The first type is a church, often located in a small building that was originally used for other (non-religious) purposes, with entrances facing directly onto the street and having a blurred boundary between the (non-religious) street and the internal (religious) space of the temple. The second type is usually located in larger buildings with a high capacity to host large numbers of worshippers. These churches are often adapted from industrial warehouses or magazines located along easily accessible urban transport corridors. In this case, there is little to no urban space for mediation between the temple and the street, leading to socialisation between the believers inside the building or, very commonly, along the road in small groups of churchgoers waiting for buses or walking. The third type has a more sophisticated architecture specifically designed for its religious function. It is located in prominent places, competing for symbolic dominance with other religions (primarily Catholic churches). In this case, the insertion is more similar to traditional Catholic temples, with a more distinct boundary between the space reserved for the faithful and the open space accessible to the general public.

In Belo Horizonte, the interplay between temples and the urban environment reveals a complex landscape of religious territoriality. As previously noted, the relationship between these religious structures and their surrounding neighbourhoods can range from being more open and permeable to more closed and impermeable. This dichotomy was observed in the central neighbourhood of Belo Horizonte, known for its thriving nightlife and entertainment scene. One traditional and non-Pentecostal Protestant temple in the area is a prime example of the more open and permeable approach. This temple actively seeks to engage with its surrounding neighbourhood, organising events and gatherings for the faithful and even inviting visitors from the neighbourhood to attend its events, including during the street Carnival period. In contrast, another conservative evangelical temple in the neighbourhood has a more closed and impermeable relationship with its surroundings, with security guards stationed at the entrance and parking lot to discourage interaction between the temple and the neighbourhood.

The territoriality of African-based religions, such as Candomble and Umbanda, is also shaped by their relationship with the urban environment. Due to the lower visibility and spatial dispersion of these temples, it can be difficult to understand the territorial dimensions of these religions. However, by taking a closer look at specific neighbourhoods with a strong tradition of African-based religions and a higher concentration of temples, such as the Concórdia neighbourhood in Belo Horizonte, we can better understand the relationship between these religions and the city. The mapping of this neighbourhood provides a more detailed picture of this relationship, offering insight into the complex landscape of religious territoriality in urban Brazil.

Visiting Concórdia

This part of the chapter aims to delve into the intricate urbanisation history of the Concórdia neighbourhood, situated in proximity to the central region of Belo Horizonte, with a focus on the peculiarities that have shaped its land use and occupation. The neighbourhood's genesis can be traced back to 1929, when it was meticulously planned and erected as a resettlement option for a working-class village that had hitherto occupied the eighth urban section of the original city plan. The decision to relocate the village was driven by the belief that its presence within the planned and affluent urban centre was incongruent with the established occupation scheme for Belo Horizonte. Intriguingly, as de Lima (2009) notes, the resettlement scheme was characterised by a unique property contract and urban legislation, which impacted subsequent years of development in the neighbourhood. The resulting occupation pattern, featuring multiple owners per lot, an abundance of popular housing typologies, and limited vacant lots, erected formidable barriers to the intervention of the real estate market. Consequently, most residents and their progeny have remained rooted in the neighbourhood, thereby endowing it with a distinctive character that continues to shape its social fabric (Figure 5.6).

This predicament was exacerbated by the protracted nature of the water supply and street paving initiatives, both of which were not realised until the 1970s and 1980s, respectively. This circumstance further perpetuated the stigmatisation of the district, ultimately leading to a diminished market value of the land. Consequently, the inhabitants of the Concórdia neighbourhood had no choice but to negotiate for urban land and increasingly populate existing lots with new structures, frequently within the same family. This history, as ascertained through interviews, played a decisive role in the district's identity and preservation of family ties, a characteristic that holds significant sway over the perception of ancestral knowledge connected to religious practices in the neighbourhood.

Moreover, the interviews revealed a palpable dichotomy between the more meticulously planned and esteemed section of the neighbourhood, with a greater presence of White residents, and the peripheral zone, characterised by a more substantial Black population. This revelation brought an additional racial dimension to the neighbourhood's identity, firmly reinforcing the structural correlation between race and social class in the country. Even to this day, the percentage of Black and Brown individuals remains higher in Concórdia when contrasted with its neighbouring districts, with the exception of regions populated by villas and slums, cementing the district's reputation as a "little Africa". This label arose not only from

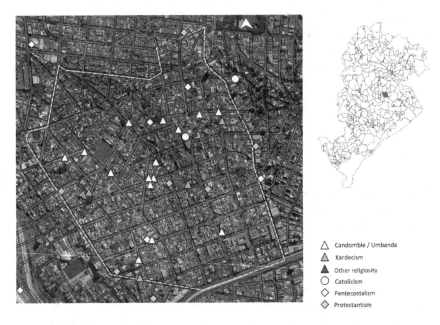

Figure 5.6 Location of temples in the Concórdia neighbourhood.

Source: Elaborated by the authors.

the substantial representation of Black residents but also from the heightened visibility of Afro-Brazilian religions.

In Concórdia, the terreiros are immediately discernible, as their presence is prominently displayed on the facades of buildings. This is a marked departure from our visits to other neighbourhoods, such as Serra, Lagoinha, Carlos Prates, Aparecida, and Área Central, where we had to exercise greater diligence in identifying them. Moreover, the African-derived religions of Concórdia are further evident in the form of religious festivities that hold sway over the neighbourhood's character and help to amplify the visibility of these faiths. Nevertheless, it is noteworthy to consider the differential treatment accorded to Christian places of worship and those of African-based religions by public authorities. While the city government has endeavoured to map the terreiros of Concórdia, none receive protection as historical or cultural heritage, and none qualify for tax exemptions related to religious activities.

Upon our arrival at México Square in Concórdia, a part of the neighbourhood renowned for its "planned and white" visual appeal, we were immediately struck by the similarities it shared with other peripheral districts in Belo Horizonte. The street layout, ambience, road network, central zones, and densely populated horizontal residential designs all seemed rather familiar. However, as we began to explore the area beyond México Square, the residential typologies became denser and more distinguishable, marked by larger backyards and more wooded areas. The terrain also became increasingly rugged, affording more extensive vistas of the neighbouring districts. While ascending a steep hill, we came across the first two religious structures, one of Afro-Brazilian derivation and the other Pentecostal. Further along, the path, as we approached the neighbourhood's periphery, we encountered a Baptist church exhibiting highly institutionalised architecture. It was out of step with its surroundings, with new cladding and minimal links to the adjoining area. Further down the commercial thoroughfare at the neighbourhood's edge, we encountered a structure affiliated with the Baptist Church of Lagoinha, an assemblage of 600 units in Brazil and beyond, with its central headquarters located in the proximate district. In the previous chapter, we discussed this particular Church and shall not explore its intricacies further. Nevertheless, it is germane to mention that this church diverges from the conventional Baptist ethos, adopting an expansionist strategy akin to that of Pentecostal groups. At the boundary between the Lagoinha and Concórdia neighbourhoods, we ascertained eight constructions affiliated with this congregation, of which three harboured moderately sized chapels.

Our discussions with the local residents suggest an absence of overt friction among the diverse religious groups in the district. The heightened visibility of African-based religions and their conspicuous presence in public arenas might imply that this neighbourhood constitutes a secure haven for these practices, even at the interface between the Lagoinha and Concórdia quarters. Concerning the interactions between African-derived religions and Christianity, our fieldwork has demonstrated that, while sporadic conflicts may surface (usually with amplified media coverage), everyday life is characterised by amity between various religious groups, even within the same kinship units. We detected no discernible obstructions or barriers in the urban milieu, even in areas marked by a specific religion.

As we meandered back towards Gabriel Passos Square, the ambience previously evoked began to re-emerge. Within the square, three ivory Gameleiras trees, imbued with potent symbolic significance for African-derived religions, engendered a space indelibly linked to these faiths within the district. Ascending a cul-de-sac that commenced with an arduous staircase, we arrived at the headquarters of the Guarda de Moçambique e Congo Treze de Maio de Nossa Senhora do Rosário and the Spiritist Center São Sebastião – Umbanda Center,[2] where Dona Isabel extended her cordial welcome to us.

Our research team had previously visited Congado, Umbanda, and Candomblé sites. Our initial impression, corroborated during our conversation, was that we received a more hospitable reception and had greater interaction than during earlier, more ceremonious visits to these institutions. In the past, we had sensed a tinge of remoteness and formality akin to that experienced in other religious establishments. One exception was our earlier sojourn to the brotherhood president and captain of the reign, Dona Neuza, where we encountered a similar geniality. However, during our meeting with Dona Isabel, a larger cohort of students, in contrast to the smaller group of researchers at Dona Neuza's residence, seemed to foster an even more cordial atmosphere and a more extensive range of themes explored. In particular, the lack of distinction between the sacred domain and the residential milieu constituted a crucial factor in our perception of the reception we received, an element emphasised by Dona Isabel and the other women we encountered in their speech. After welcoming us with embraces and individual greetings, Dona Isabel underscored the importance of "looking after those who open their home", a hallowed space safeguarded by her faith.

Throughout our protracted dialogue with Dona Isabel, she underscored her family and religious group's ancestry, history, and themes related to racism and inequality. She specifically referenced the urbanisation history of Concórdia and the discrepancies between Praça México and Praça Gabriel Passos, in addition to the significance of religious events for the visibility and territorialisation of religion. When asked about the role of urban planning and what students should contemplate in this profession, Dona Isabel emphasised the need to cherish the sacred and comprehend that what is sacred is "holy for each one". She then recounted the tale of Paul of Tarsus and his opposition to the structures he helped erect, providing a pointed critique of planning and its impact on the city.

According to Dona Isabel (Gasparino & Torres, 2021), the formation of territory through the visibility of religious events has a momentous import in terms of the acceptance and proliferation of religion. She expounded, stating that "when we venture out into the streets, into the urban space, we expose ourselves to the gaze of those others who were not anticipating our appearance and who have no notion of what Reinado, Rosário, or Congado is, and we can attract more people". Nevertheless, there is also a perception of a transcendent objective, as "when these potent individuals pass by, with an awe-inspiring and primordial vigour, each stride they take in this urban space transforms energies: the negative into positive, the bad into good, and negative thoughts into good thoughts". Dona Isabel observed that the scale of the street serves a distinct purpose than the domestic cult, elucidating that "the domestic cult accomplishes what it must, but when we depart our home and traverse the streets, we transcend. It is a healing energy that extends far beyond and has far greater reach".

Concluding remarks

In this chapter, we have scrutinised the spatial allocation of buildings linked to various religious groups, situating them within the larger historical, urban, social, and economic context. This materialist perspective is certainly valuable in comprehending the religious landscape of Belo Horizonte. However, we argue that the final section of the chapter, which scrutinises Concordia, offers a much-needed micro-urban perspective, enabling us to articulate our main argument more fully: for many inhabitants of Brazilian cities, religion is intimately interwoven with everyday mundane practices, and the religious aspects of their lives are only partly spatial. Instead, religion is contextualised "horizontally" by the surrounding spatial practices of other communities and "vertically" by the histories and genealogies of specific phenomena. Droogers' three dimensions of analysis – internal, external, and transcendental – prove particularly advantageous in examining religious practices at the district level.

Notes

1 Mapping carried out in four Brazilian metropolitan regions (Belo Horizonte, Belém, Recife, and Porto Alegre) by the Ministry of Social Development and Combating Hunger (MDS), United Nations Educational, Scientific and Cultural Organization (UNESCO), Secretariat of Policies for the Promotion of Racial Equality (SEPPIR), and the Palmares Cultural Foundation (FCP). The project aimed to identify who they are, where they are located, what are their main activities, how the land situation, infrastructure, and other sociocultural and demographic aspects of the terraces are. Go to the project website for more www.mapeandoaxe.org.br.
2 Guarda de Moçambique e Congo Treze de Maio de Nossa Senhora do Rosário and the Spiritist Center São Sebastião – Umbanda Center.

References

Alves, J. E., Cavenaghi, S., Barros, L. F., & Carvalho, A. A. D. (2017). Distribuição espacial da transição religiosa no Brasil1. *Tempo social, 29*, 215–242.

Arenari, B. (2015). Pentecostalism as religion of periphery. https://doi.org/10.18452/17182.

Beccneri, L. B., de Farias, L. A. C., & Chiroma, L. (2019). Transição religiosa e divisão do espaço urbano: uma análise da década de 2000. *PLURA, Revista de Estudos de Religião/ PLURA, Journal for the Study of Religion, 10*(2), 185–206.

de Lima, J. M. F. (2009). *Bairro Concórdia em Belo Horizonte: entrave ou oportuni dade à cidade-negócio?.* [Master thesis]. https://repositorio.ufmg.br/handle/1843/RAAO-7YHHCZ

Gasparino, I. C.; Torres, J. (2021). O Reino nas ruas. Piseagrama, Belo Horizonte, no. 15, pp. 2–9.

Gil Filho, S. F. (2006). Estruturas da territorialidade católica no Brasil. Revista Scripta Nova, Revista Eletrônica de Geografia y Ciencias Sociales. Universidad de Barcelona. v.X, n.205.

Gil Filho, S. F. (2012). Espaço Sagrado: estudos em geografia da religião. Curitiba: InterSaberes.

Machado, M. D. D. C. (1996). *Carismáticos e pentecostais: adesão religiosa na esfera familiar.* Campinas, SP: Autores Associados; São Paulo, SP: Anpocs.

Mafra, C. (2011). Dossiê: O problema da formação do "cinturão pentecostal" em uma metrópole da América do Sul. *Interseções: Revista de Estudos Interdisciplinares, 13*(1), 136–152.

Moura, J. L. P. (2019). The geography of the sacred in ketu candomblé terreiros. *Revista África e Africanidades,* Ano XI/29

Oliveira, H. C. M. (2012). Espaço e religião, sagrado e profano: uma contribuição para a geografia da religião do movimento pentecostal. *Caderno Prudentino de Geografia, 2*(34), 135–158.

Rocha, C. (2019). "God is in control": Middle-class Pentecostalism and international student migration. *Journal of Contemporary Religion, 34*(1), 21–37.

Spyer, J. (2020). *O Povo de Deus.* Nordersredt: BOD GmbH DE.

6   Temple, street, home, and nature

Look, it's... I used to guide people to pray at home and search in the family vein. Those predisposed to go to the mount at certain empty times, if they wanted to go, would go to seek refuge and spiritual reinforcement. But the most sacred thing was the home, right, the person's home. There was no other space because the only space we have as faith that we have as sacred is either the temple or the mount when we go to make purpose or our house, which is our home, our family home, outside of that... I at least don't see another place as sacred.

(White man, 49 years old, a pastor in his own church.
This interview was conducted in April 2021)

Four places to look for God(s)

This chapter delves into the performative aspects of various spatial categories: home, temple, street, and nature. Through the lens of Brazilian authors such as Roberto daMatta (1985), Angelo Serpa (2019), and Marcelo Lopes de Souza (2013), as well as the perspectives of Brazilian believers, we aim to unravel the meanings attributed to the places where God, god-like entities, or spiritual forces are perceived. At the heart of this discussion is that each of these categories constitutes a distinct set of imaginaries shaped by the beliefs and practices of those who inhabit them. In this sense, the authors invited to contribute to this chapter provide a unique and insightful perspective on the performative aspects of space.

It is worth noting that the theoretical framework underpinning this discussion is rooted in the work of French philosophers Henri Lefebvre and Pierre Bourdieu, whose influence on the authors of this book and on the production of knowledge in Brazil is undeniable. Despite our efforts to pursue a framework informed by Brazilian (non-Western) authors, the contributions of Lefebvre and Bourdieu to our understanding of space and our own academic formation cannot be overstated. Still, in this chapter, we invite the reader to engage with the performative aspects of different spatial categories and to comprehend their meanings mainly through the perspectives of Brazilian authors and believers.

The appropriation of various spaces within the city is shaped by their unique configurations and characteristics (Santos et al., 1985). In the case of Brazil, its rich cultural diversity is a defining feature, and as Roberto daMatta (1985) suggests, the

DOI: 10.4324/9781003248019-6

street and home have been central to attempts to define common themes in Brazilian culture, essentially creating a national archetype.

However, it must be acknowledged that this constructed archetype is exclusionary, focusing primarily on individuals belonging to popular classes residing in urban or metropolitan areas and historically marginalised by colonisation. This archetype fails to fully encompass the experiences of indigenous and other native groups and those belonging to communities such as quilombos and caiçaras,[1] who may have their own unique customs and traditions. This is particularly relevant in Belo Horizonte, where these groups are not present in significant numbers.

For most individuals, the home and street form the foundation of daily life and are the spaces where we spend most of our time interacting with the world. Even during times of crisis, such as a pandemic, we inevitably traverse these spaces on a daily basis. The home and street exist in a relationship of opposition, with the former representing the private sphere and the latter representing the public realm.

It is important to note that these categories encompass more than just physical locations. They also include various social practices and ethical norms shaping our understanding of the world. As Roberto daMatta highlights, the home and street are not simply geographic spaces but rather social categories that influence our perception and interpretation of reality.

In our interviews, a distinct pattern emerged in how the believers viewed their environs as spaces where the divine could take on various incarnations. For Christians, especially Pentecostals, the entire abode was considered under divine protection. Yet, they felt a deeper connection in more private quarters – a bathroom, bedroom, or even a car or bus. These Pentecostals felt a connection to the Holy, even in solitude, but that connection was amplified in the presence of their pastor, whether in a church or through televised prayers. This leads to the realisation that there is no such thing as a "sacred space" per se; instead, it's a matter of varying degrees of connection to the divine. While space plays a role, it's just one among many factors.

The street, in contrast, is a melting pot of diversity and danger, a place where people interact with one another and their leaders and where political and social events unfold. It's the polar opposite of the home, which is supposed to be a haven of refuge and repose, where the outside world's demands are left at the door. But, as the COVID-19 pandemic has shown us, this isn't always the case. The home is also a hub of family life and social gatherings, governed by a labyrinthine set of norms and values.

These distinctions between the home and the street, as well as other public spaces, are shaped by cultural norms, resulting in a social hierarchy. The home serves as the nucleus of familial relationships and the street and other public spaces serve as the crucible of labour and politics. The aforementioned authors did not explore the concepts of temple and nature, but we propose to introduce these categories and examine them within the framework of religious imagination, viewing them as "gateways" to transcendence.

For both Pentecostals and Condomblecistas, the temple represents a space that differs from the public sphere and can thus be considered a separate category. On the other hand, both groups view nature as untainted, yet they also manipulate it for sacred purposes – for example, by incorporating plants into their personal altars. To simplify this category, we would say that the temple may be understood as the junction between home and street, while nature is considered an undisturbed product of the divine. As always, any type of categorisation is problematic and could be questioned. As we will discuss further, these categories have slightly different meanings for Pentecostals, different for Candomblecistas and different in academic discourses.

Through examining these differences and similarities, we can better understand how communities attribute meaning to spaces and appropriate them. There is a relationship between the definition of space and the values associated with it. For example, the distinction between "street" and "home" can reflect two sets of norms and behaviours, creating two distinct types of Brazilian society: home-Brazilians and street-Brazilians. In the same vein, we posit the existence of temple-Brazilians and nature-Brazilians.

Perception and social representation of space – what do we know about these places?

As the reader is already well aware, this chapter and the entire book are influenced by Henri Lefebvre's and Pierre Bourdieu's theories. Although we anticipate that most readers are already familiar with these ideas, we will briefly touch on Lefevbre's three dimensions of space (conceived, perceived, and represented) and delve into Bourdieu's concept of habitus. We aim to examine how religion and faith (secondary habitus) can alter how space is perceived, conceived, and represented and to observe how these changes manifest through the various elements and actors that comprise a given space. This is of great significance as it highlights the larger structures that shape modern life and how they are reflected in the creation and organisation of space.

Lefebvre's ideas have been updated and refined by leading experts in Brazil who focus on analysing space. In the late 1990s, Brazilian Marxist-Lefebvrian scholars, such as Angelo Serpa, Danilo Volochko or César Simoni Santos, formed a robust theoretical and methodological foundation that continues to shape the study of geography to this day. According to Lefebvre, space symbolises the concurrent and simultaneous aspects of social reality, while time symbolises the historical progression of the production process. As social phenomena, time and space must be analysed in the unique context of the societies in which they exist.

The production of space theory is composed of three key elements: spatial practices, representation of space, and spaces of representation. The first aspect pertains to the space itself and the actions that occur within it, including both physical and social practices. These practices are shaped by human and natural forces and can encompass transportation, construction, and land ownership. The second aspect, representation of space, refers to the way in which we describe and depict space through languages, such as in maps or photographs. The third aspect, spaces of representation, highlights the way in which individuals make sense of and understand space. To bring these elements to life, we can consider a person's daily commute from home to work or temple as an example of spatial practice. The maps we create to represent this space would be an example of representation of space, while the categories we use to describe it (such as "temple", "tree", or "nature") are examples of spaces of representation.

This third dimension, which we find particularly fascinating, delves into the way in which we make sense of the world around us – be it through natural elements like trees and mountains, man-made structures like buildings and monuments, or a combination of both. By comprehending this third dimension, we gain insight into how spaces can evoke social norms and rules (in our case, spaces that may bring a believer closer to the divine) and serve as a social experience. Perceived space is a sensory and subjective phenomenon encompassing the full range of our experiences. Lefebvre posits that the irrational is a social reality that forms the backbone of the social imagination. Yet, in order to be perceived, spaces must first be conceived – meaning there must be a discourse that references space, such as a map, a picture, or a description – and they must also be lived, meaning they must have substance, not just form.

In the words of Brazilian geographer Angelo Serpa, the geography of social representation is marked by the overlapping of experience and appropriation of urban spaces. This overlap, Serpa argues, results from the interplay between lived space and the meanings attached to it. Thus, those who hold sway over spaces of representation – the state, artists, or temple owners – can impose their worldview and social imagination upon these spaces. This often leads to conflict and tension in creating spaces meant to represent a certain meaning. To better understand this dynamic, one might consider the role of architecture in shaping the social imagination. By examining the most prominent and "assertive" forms of buildings, we can gain insight into the ways in which they serve as powerful symbols in the collective imagination. As we have discussed in another text,[2] the main church of Igreja Universal do Reino de Deus (IURD) in Belo Horizonte could be seen as "... a massive icon of a prosperity gospel", which is present not only in the physical space of the city but also "... the scenographies and general visual references used in the IURD's soap operas are based on the visual and spatial qualities present in IURD's churches".

Bourdieu's ideas offer a roadmap for delving into the intricacies of lived religion. At the heart of these ideas are the concepts of habitus, field, capital, social reproduction, and domination. Religious symbols and codes, however powerful they seem, don't simply exist in a vacuum. They must be "activated" by those who practice the religion. Here, the idea of habitus – a system of ingrained tendencies that dictate an individual's actions within different fields – becomes critical. The habitus represents the real-time impact of an individual's or group's past, memories, and history on their daily lives. For example, the religious practices of one's childhood can have a lasting impact on the habits and actions of adulthood.

In the temple, we find a site of worship and community, a place where believers gather to fulfil their religious obligations and form bonds on a spiritual plane. This temple is the dwelling of the divine, a space of religious community where the notion of "the other" takes on a distinct meaning compared to that in the secular street or the private and personal space of the home. Some interviewees reported frequenting the temple multiple times weekly and engaging in activities beyond just the rituals. For some, the temple serves as a place for work and study; for others, it acts as a social hub, functioning as an "extended family". The temple holds a significant place in the schedules of believers and is a vital aspect of their daily lives. It can also be seen as a space where the dividing lines between the home and the street intersect, as it is both a public entity, accessible to all and promoting interaction with outsiders and a place for establishing customs and rules similar to those of the home, fostering closeness and a feeling of belonging among family and peers. The street, with its practical and functional purpose, allows for the coexistence of religious and secular activities, while the home serves as a refuge for individual spiritual pursuits and familial intimacy. Beyond the confines of these human-constructed spaces lies the boundless expanse of nature, an existence beyond the reach of human creation.

These four (spatial) categories – the temple, the street, the home, and nature – are each defined by their own set of ethics, forged by their particular features and the beliefs of those who occupy them. Even within one religion, these spaces may be perceived with varying degrees of reverence, with some devotees embracing the sanctity of other temples and others regarding certain sites as profane. Take, for instance, the Candomble follower who views all temples as sacred, as opposed to the Pentecostal who considers African religious temples evil. And for a Candomble practitioner reared in a Catholic household, the notion of what is sacred may shift depending on the specific space – the terreiro, or Candomble temple, may be seen as requiring greater respect due to its dual function as a place of worship and a person's home, while the Church is regarded solely as a house of God.

A shared belief system and common practices within a temple engender a powerful sense of community and belonging among its congregants. In effect, it provides a surrogate family of sorts comprised of individuals who not only support one another but also share a deep understanding of each other's spiritual pursuits. Nevertheless, in societies where multiple religions coexist, tensions and conflicts may arise. The church or temple thus serves not only as a "sacred" space but also as a social one, and the bonds forged therein frequently transcend the physical limits of the building itself. Indeed, as our interviewee E4 (Male. Pastor of the Church/Computer Technician, 49 years old, Neopentecostal) so astutely observed:

"The Church has its ups and downs, right? Sometimes it is full, sometimes it holds 60 people, sometimes it holds 80 people, and sometimes it holds 10, 12, so it is relative (…) Some are in the Church not for their lives, for their souls or for the faith they profess, but for the affinity of friendship right, "ah, I'm here because I'm friends with that person, I'm here because of that person," so when that person doesn't go she won't go. So we have this; the Church lives on this, unfortunately. The Church exists because of the human being; that's what I say."

Each religion's distinctive values and ethical codes are inextricably bound up with the internal hierarchy and structure of the faith itself. But the physical architecture and use of religious spaces and buildings also play a crucial role in shaping the faithful's perception of the rules and values of the religion. In other words, a religion's material embodiment can profoundly impact how it is apprehended and experienced by its adherents.

However, the notion of divine presence is by no means limited to the confines of temples and religious edifices. On the contrary, the specific sites believed to house the divine can vary greatly across religions. For instance, practitioners of Candomble subscribe to the view that their deities or Orixás can be found ubiquitously – in public spaces, the natural world, and even in the products of human technology. As we have previously established, the concepts of sacred and profane are pretty blurred in Candomble. One of our interviewees, a Candomble practitioner, underscored that the Orixás are omnipresent not only because they manifest in all aspects of nature and human creation but also because believers can assimilate the Orixás into their own essence during ceremonies at the terreiro. In Candomble, it is maintained that the Orixás take up residence "within" the believer following their initiation into the religion, so their presence is felt by believers wherever they may be. By contrast, Pentecostals frequently feel a more potent connection to the divine in certain physical locations; however, the Holy Spirit also can work through the body of the believer.

An intriguing parallel between Candomble and Pentecostalism is their shared conviction that their respective deities can be encountered even in the most mundane of surroundings, including one's own home and bedroom. It is noteworthy that all of the interviewees, irrespective of their particular faith, expounded on the omniscient nature of their God/desses. However, the contrast between the two religions rests on how their omnipresence is perceived. For instance, while Pentecostals may experience a more potent connection to the divine in quiet spaces, nature, and temples, Candomble practitioners contend that the presence of the Orixás is discernible in every nook and cranny of the world and in all things. We propose that this is the fundamental distinction: Pentecostals evoke more the idea of a connection with God, while Candomblecistas speak rather of the presence of Orixas.

According to our Candomble interviewees, every component of the universe – every person, object, and substance, whether "good" or "bad" – constitutes a fragment of the larger whole, Oludumaré. One interviewee revealed that they felt attuned to the divine even while visiting the market, owing to the food's particular significance as both a life-sustaining necessity and a product of the Orixás. Another interviewee recounted visualising the Orixás as possessing distinct faces or being embodied by natural elements, such as seeing Exú in the streets (associated with movement), Iemanjá in water, or Oxóssi in knowledge or forests. Consequently, practitioners of Candomble perceive all facets of nature and human creations as inextricably bound up with the Orixás.

In contrast to Candomble's perception of the interconnectedness of all things, Pentecostal Christians frequently conceive nature as a divinely created domain that remains unsullied by human intervention. For Pentecostal interviewees, the presence or absence of human-made structures can either fortify or weaken the divine presence. Consider, for instance, the notion of a temple. These structures, constructed by humans for religious devotion, are regarded as sacred spaces. However, in the midst of the frenzied urban milieu, many other edifices may be deemed profane. Yet, for adherents of Candomble, the human element is mostly irrelevant. By contrast, Pentecostals frequently view nature as the antithesis of the built environment – a locale where one can encounter the home, the street, and even the temple. Within this paradigm, nature is revered precisely because it remains unsullied by human influence. Consequently, the natural world is susceptible to being desacralised by human activities. This view can lead to calamitous outcomes in Brazil, where Pentecostalism has forged (at least in part) a political alliance with agribusiness.

It is evident that the interconnections among the four categories – nature, the home, the street, and the temple – are intricate and frequently contentious. The built environment, replete with its artificial structures and cities, often contrasts nature's untrammelled, primal force. However, for many of the adherents we interviewed, it is in the natural world where they experience the most profound communion with the divine, a relationship that is noticeably absent in the other three categories. Hence, we discern that the relationship between the sacred and the profane, between the natural and the constructed, is multifarious and marked by paradoxes and tensions. This dynamic necessitates constantly reevaluating our comprehension of what it means to be connected to something greater than ourselves.

In light of the preceding discussion, conducting a more structured comparison between Candomble's and Pentecostal perspectives on the four categories is worthwhile. For Candomble, the interconnections between nature, the home, the street, and the temple are inseparable and of equal value. The natural world is imbued with the presence of the Orixás, and human-made structures are seen as a fundamental component of existence rather than a separate category. In turn, the home, street, and temple are all potential sites of Orixá manifestation and are thus held in similar esteem.

By contrast, Pentecostals distinguish between the sacred and profane, with the natural world often viewed as a realm of the divine, untainted by human intervention. As such, nature is deemed the most conducive location for experiencing the divine presence. Human-made structures, including the home and the street, are considered profane compared to sacred temples. The temple is a space explicitly set apart for religious practices and is therefore accorded a higher degree of holiness than the other categories.

These differing perspectives indicate that the nature-constructed, sacred-profane dynamic is by no means uniform across different religions. Candomble's and Pentecostal's approaches reveal a complex interplay of values and beliefs, with the distinction between sacred and profane proving to be a particularly contentious issue.

Home

As alluded to earlier, the notion of the home carries significant social weight in the Brazilian context. This is owing to the fact that the domestic space embodies a set of norms and values that shape the conduct of those who dwell within it. Additionally, the house is closely bound up with the concept of property, a highly contentious issue in Brazil, where numerous citizens lack their own homes. The movement for agrarian reform among landless workers has been in operation for decades. Nevertheless, the house is more than just a physical structure; it is also a socio-economic entity that belongs to someone. Consequently, the regulations that govern behaviour within one's home may vary from those that apply in the home of another. DaMatta (1985) underscores the distinction between the house and the street, even though it is important to note that these two spaces are not always mutually exclusive. Many people may dwell on the street due to a dearth of housing, while others may perform work (activity associated with the street) from within their homes. This underscores the fluidity and complexity of the relationships between these different spaces.

For daMatta, the house represents a private and intimate space that assumes a diverse range of connotations contingent on the context – it could allude to one's hometown, a bedroom, or any location one considers home. This connection between space and social order is further complicated by the presence of a "house code" predicated on notions of family, friendship, and loyalty, as opposed to the "street code" that governs behaviour outside the home. Within the Brazilian context, the house has traditionally been viewed as a site of tranquillity, respite, and graciousness. Nevertheless, when we examine the influence of religion on the house's signification, we discover that this space acquires even more subjective attributes.

Irrespective of their specific faith, our interviewees unanimously attested to their belief in the sacredness of their homes. For practitioners of Candomble, this process of sacralisation involved the infusion of religious elements into the physical space of the home, such as plants, art, and artefacts that symbolise the Orixás. In contrast, for Pentecostals, the sacralisation of the home was linked to a sense of serenity and individual devotion, demonstrated through the construction of altars and the presence of the Bible.

This subtle disparity in attitudes towards the home as a locus of religious observance was brought into sharp relief by the events of 2020 when the COVID-19 pandemic mandated social isolation and limited access to temples for many individuals. During this period, people were obliged to rely on their homes as the primary venue for religious worship. When queried about how they sustained their faith during this time of seclusion, the two groups evinced divergent reactions. While Pentecostals could maintain their religious practices through virtual services and social media exchanges, all Candomble practitioners we interviewed affirmed that practising their faith remotely was impossible. Instead, they turned to study as a way to stay connected to their religion. This indicates that while the home may be sacred for both groups, it may not necessarily be a site of religious practice for all in the same way.

An aspect of religious practice that is unique to Pentecostal groups is the extensive employment of television and other media as a medium of mediation. The Universal Church of the Kingdom of God (IURD) is a particularly striking example of this phenomenon, as it possesses the third-largest television network in Brazil and numerous radio stations, newspapers, and publishing houses. This pervasive media presence allows the IURD to disseminate its teachings to a vast audience and facilitates its followers' access to the Gospel and prayer, obviating the need to physically attend a temple to engage in these rituals. Furthermore, the IURD's teachings are a reference point for other Pentecostal congregants.

Street

The street is another pivotal category of analysis, symbolising the external world beyond the home. The interviews and daMatta's contributions indicate that the associations between individuals and public space vary considerably. Space and social order are inextricably linked, and this dynamic is particularly pronounced in the streets. The space itself is shaped by the interactions between individuals, becoming part of their comprehension and imagination of those interactions. Upon analysing the various groups, it becomes clear that there is a distinct perception of the street in contrast to the home, based on each group's values and morals. The constructed space itself carries the weight of these perceptions, influenced by the belief systems conveyed through religious doctrine. Candomble practitioners contend that they can sense the presence of the Orixás wherever they may be. Nevertheless, regarding public spaces, especially those linked with movement, physical pleasure, and public rituals, the figure of Exu is frequently cited by believers. In Candomble beliefs, Exu is responsible for movement and communication and is characterised by interviewees as "responsible for things to happen; he is responsible for the chaos". Offerings are regularly made to Exu at crossroads on the streets since these are deemed to represent encounters – that is, the intersection between the earthly and spiritual worlds, which is mediated by Exu. Although Candomble practitioners maintain that there are no profane spaces, many of them also mention that the energies of Exu can be, at times, too intense, leading them to avoid places such as bars or extremely crowded stores. Even so, such spaces are still regarded as being imbued with sacred energy. While examining the street, it is noteworthy that Exu is often stereotyped as the devil by Christians (mainly Pentecostals), resulting in the destruction or denunciation of Exu's offerings at the crossroads.

Pentecostals perceive the street as a site of profanity. Worldly celebrations and festivals are events to be shunned, as the street is viewed as a location where interactions with those who do not follow their religion may take place. As a result, it is deemed safer to avoid such interactions.

Temple

The temple occupies a unique space in religious practice for both Pentecostals and Candomble practitioners, serving as a site of worship, socialisation, and communal activities. As a distinct category of place, the temple is seen as a sacred space, occupying what daMatta refers to as the "other world" – a realm that transcends the rules of the home and the street. For Pentecostals, the temple is the house of God and a place for socialising with other church members. It also provides a space for addressing personal struggles and engaging in philanthropic efforts. In contrast, Candomble practitioners view the temple as a sacred space for initiating new members and conducting rituals, often involving invoking orixás. These rituals are central to religion, as they allow individuals to connect with their spiritual selves and the wider community. Despite their differing perspectives on the temple, both groups demonstrate the importance of sacred space in religious practice and the ways in which it can serve as a site for social interaction and communal activities beyond the realm of the home and the street.

Candomble practitioners view the temple as the house of someone else (the Father and Mother of the house: "pai de santo" and "mae de santo"), placing significant emphasis on the intentions of those who occupy them to imbue them with sacredness. Consequently, the temple is a space that must be respected, as it is considered the priests' home. However, the purpose of this space goes beyond its religious function, serving as a refuge for believers from the challenges of everyday life. The initiation of believers into the religion takes place within the temple, a process that involves separating oneself from the outside world and participating in the maintenance and study of the secrets of the religion. This initiation creates a sense of belonging to the house, which is developed through an emotional bond akin to that of a home. The temple is also regarded as a place to prepare and shape believers for the outside world.

A song by a group of Candomble musicians, popular in Brazil due to the participation of two well-known Brazilian pop singers, attempts to capture the essence of the terreiro. In the interviews, believers often mentioned this song when discussing their relationship with the terreiro. The first part of the song includes the following lyrics:

> O terreiro é uma casa acolhedora, de irmãos/Onde aqui a gente perde os títulos lá fora/São todos aqui meu pai, minha mãe, meu irmão/É uma família unida/Essa casa, esse terreiro, como queira nominar/É um grande útero, onde cabe todos os seus filhos/E todos encontram aconchego, respeito, carinho/E, quando necessário o apoio
>
> The terreiro is a welcoming home, of brothers/Where here we lose titles abroad/They are all here my father, my mother, my brother/It's a united family/This house, this terreiro, whatever you want to name/It's a large uterus, where all your children fit/And everyone finds warmth, respect, affection/And, when necessary, support.

The terreiro is considered an extension of the home by Candomble practitioners. It is a physical space located outside the home. Yet, it embodies characteristics that are typically associated with the home, such as creating strong bonds similar to those found within a family. Believers even describe the terreiro as a "uterus", a place of birth and creation, where initiates are introduced to the world of Candomble. As a result, the terreiro is a vital space for fostering a sense of belonging and community within the religion.

Nature

Nature is a category of utmost importance in our analysis, and unlike the home, street, or temple, it is often regarded as untouched by humanity. However, this view can be contested because the relationship between humans and nature is deeply intertwined. As Brazilian geographer Milton Santos (1994) has argued, we now live in a technical-scientific-informational environment where the distinction between the natural and the artificial is no longer clear. Today, nature cannot be separated from technique or science, as the discourse on Anthropocene/Capitalocene has made clear (Malm, 2016; Moore, 2015,). Human beings undoubtedly touch, transform, and work within nature, and this has significant implications for how we perceive it.

As demonstrated through our interviews, both the neo-Pentecostals and Candomble practitioners view nature as a manifestation of the divine. For the former, nature is seen as the untouched work of God, while for the latter, nature represents the Orixás. These powerful spirits or deities are associated with different elements of the natural world and are considered to represent the supreme being, Olorum. The orixás are not only present in nature but also in the built environment and technological products, according to the Candomblecistas, who believe that everything originates from nature. As such, nature serves as a gateway to the Divine for Candomblecistas, as they believe that the traces of the divine can be found everywhere and that the faithful can connect with the Holy through external phenomena at any time and in any place.

In contrast, for the Pentecostals, the only true connection with God is through the individual believer. Nonetheless, nature still holds a sacred significance for both groups, as it serves as a place of worship and connection with divine forces.

Final remarks: Too close and too far away

It is imperative to note that an essential albeit nuanced contrast exists in the perception of the sacred and profane between two groups, which is expressed in the materiality of their respective convictions. The concepts of the sacred and profane are not static or fixed, but instead, they are fashioned and imagined based on the knowledge and beliefs of each adherent. Therefore, it is impossible to neatly distinguish these notions across (spatial) categories such as the home, temple, street, and nature. Nevertheless, for Candomble practitioners, the notion of the profane is a foreign one: spaces, actions, and objects can all be regarded as sacred to varying degrees, and the lack of sacralisation does not equate to being profane. One could claim that there are no correct or incorrect practices in Candomble, and as such, there are no profane spaces or practices. However, this also implies that nothing is necessarily sacred in the way that Christians perceive sacredness.

In contrast, Pentecostals acknowledge the existence of the profane, which is not a neutral notion but rather carries connotations of "profanity". Thus, for Pentecostals, the concept of the profane is often associated with evil. This observation clarifies why there is an element of urgency in Pentecostal endeavours to construct a "sacred infrastructure", as their religious project aspires to eliminate malevolent forces from the world at large.

This distinction highlights the diverse ways in which these two groups construct their understanding of the world and their relationship with the divine.

Notes

1 Quilombos are communities formed in Brazil as a result of territorial, social, and cultural resistance. They operate based on the culture and traditions of their inhabitants, who are typically former enslaved black people who fled in search of freedom. These communities have preserved their African cosmovision and family ties, organising themselves as autonomous entities that still exist today.

   Caiçaras, on the other hand, are the traditional coastal populations of the South and Southeast regions of Brazil. They are a result of the mixing of indigenous people, Portuguese settlers, and former slaves, and are considered one of the last visible remnants of the formation of the Brazilian people.
2 See Cavalcanti de Arruda et al. (2022).

References

Cavalcanti de Arruda, G. A. A., de Freitas, D. M., Soares Lima, C. M., Nawratek, K., & Miranda Pataro, B. (2022). The production of knowledge through religious and social media infrastructure: World making practices among Brazilian Pentecostals. *Popular Communication, 20*(3), 208–221.

DaMatta, R. Roberto. (1985). A Casa ea Rua–Espaço, Cidadania, Mulher e Morte no Brasil.

Malm, A. (2016). *Fossil capital: The rise of steam power and the roots of global warming.* Brooklyn, NY: Verso Books.

Moore, J. (2015). *Capitalism in the web of life: Ecology and the accumulation of capital.* Brooklyn, NY: Verso Books.

Moore, J. W. (Ed.). (2016). *Anthropocene or capitalocene? Nature, history, and the crisis of capitalism.* Binghamton, NY: Pm Press.

Santos, M. (1994). Técnica, espaço, tempo: globalização e meio técnico-científico informacional.

Santos, C. N. F. D., Vogel, A., & Mello, M. A. (1985). *Quando a rua vira casa.* São Paulo: Projeto.

Serpa, A. (2019). *Por uma geografia dos espaços vividos: geografia e fenomenologia.* São Paulo: Editora Contexto.

Souza, M. L. D. (2013). Os conceitos fundamentais da pesquisa sócio-espacial.

7 Interlude

Religion and music (Daniel's story)

When you are near a drum, an ancient food, or a geographic place that messes with you and you, out of ignorance or mental slavery, accept or repudiate, that is self-evident. This is because when the drums play, when nature touches the human being through the mother, through the mitochondrial DNA, which is Africa, there is no escape for anyone. Now, are you going to accept this or not?

(Camilo, black man and artist. This interview was conducted in February 2023)

I

I had minimal religious training. In the house where I spent my childhood, I lived with a large family of uncles, grandparents, and cousins, each one of them Catholic, Protestant, and kardecist. All were non-practitioners of their respective religions; they believed in God, attended specific masses, prayed individually, and recommended that their children make their first communion. A few conflicts appeared in the debates about the importance of having a religion or not or respecting each one's faith. There were no practitioners of African religions in the family, although I remember occasional reports of consultations with mothers of saints and visits to terreiros. The new religiosities frequented the house, but in a more superficial way, through brief incursions by some into oriental meditation or astrology.

Discussions about music, on the other hand, were commonplace and heated. It was discussed, for long hours, what should be listened to or avoided, and how each artist related to politics, behaviour, and, to a large extent, with what we could call the evolutionary line of popular song in Brazil, a term that I heard used by Caetano Veloso.[1] That's why – greatly influenced by my family's taste – I started to listen, research, discuss, and play every day of my life the musical genre that in Brazil is usually called MPB (Brazilian Popular Music).

Today it bothers me that the debates between religion and music didn't mix at home, except for occasional annoyance when artists were too aggressive or devoted to a particular religion. When the presence of religion was more evident in music, there was a tendency to convert it into a cultural heritage or recognise that faith concerning the Christian reference where "everything is God, regardless of the name given". The fact is that failing to inherit the religion of my relatives, music was the primary interface I kept with my family and the main device that

DOI: 10.4324/9781003248019-7

instigated, illustrated, and preserved the ethical and moral content of the debates that formed my personality and vision of the world. A good way of illustrating the importance of the song is "Balada de Tim Bernardes" (Bernardes, 2022), which does not deal directly with religion. Still, it presents a path that starts from his grandmother's relationship with "issues about witchcraft" and with the "mysteries of the planet and nature's gift of transforming itself" for a sensitive reflection on the central role of songs in his formation. The musical aesthetics of the music and the video are exactly what this text tries to show.

I avoid the commonplace of saying that I find my spirituality in the arts, especially in music. This is partly true, but I believe there are more interfaces to be explored in how music and religion dialogue in Brazil. There is a vast amount of material on the relationship between music and religion, but it is not the purpose of this text to use this material. The objective here is to revisit from memory some moments of the interface between music and religion in my life. With today's eyes, trained by the research that generated this book, I try to understand how religious themes in Brazilian songs affected my perception of the world and how it prepared me for a more natural and tolerant contact with religious content and practices.

I still do not practice religion, although I am increasingly interested in how religion structures our relationship with the world. I will emphasise some songs and, despite having made an effort to go through the composers' discography in search of a more rigorous mapping of the theme in each artist, I privileged the songs that I listened to more recurrently or that impacted more explicitly. On this point, I realise today that a closer look at the theme reveals more subtle and constant presences in the work of some of the artists, where religion, more than a theme, acts as the structuring axis of their work and behaviour, as I will name throughout the text. A second caveat stems from playing a musical instrument, which causes attention to the lyrics to be neglected due to other musical elements. If, on the one hand, this generates a lack of attention to the content, on the other hand, it allows me to identify the more subtle presence of elements of religion and how they were articulated with the musical structure of each song, an essential aspect for the effectiveness of the orality contained in the songs. Playing an instrument also forces you to create a repertoire, a limited set of songs constantly revisited and turned over, expanding the intensity of the relationship with the apprehension of these songs' musical and poetic content.

In this sense, much of my memory and initial hypotheses dealt with in this text were revisited and influenced by my brief conversation with Camilo Gan[2] in February 2023. I left the chat more attentive to the understanding that orality in African-based religion is not limited to saying certain content. Still, it contains a sophisticated ancestral technology capable of linking axè to the meaning of what is intended to be conveyed. My initial understanding of this speech, greatly influenced by my poetic and musical references, referred me to the structure of the "sambas de roda" and, more particularly, to the work of Dorival Caymmi, two concerns linked to Bahia, the first capital of Brazil and the place of the most significant presence of religions of African origin. I will talk about Caymmi later on.

"Samba de Roda" is a musical form that emerged in Bahia in the 17th century, closely related to traditional festivals and the cult of the Orixás. In my view, the song Reconvexo (Veloso, 1989) presents a reading of how samba de roda is directly related to religion. Composed by Caetano Veloso to be interpreted by his sister Maria Bethânia, both born in the Recôncavo region of Bahia, it lists situations where it would be possible to see the presence of the sacred, although with the warning: "your eye looks at me, but it cannot reach me" (seu olho me olha, mas não me pode alcançar). The song, in which the lyrical self would be this presence of the sacred, reinforces the ancestry, matriarchy, and territorial specificity of Bahia in the passage "I am the shadow of the voice of the matriarch of Roma Negra" (eu sou sombra da voz da matriarca da Roma Negra). The sound and structure of the song are very close to traditional "sambas de roda" and mixes elements internal to the Recôncavo culture with foreign and contemporary features, suggesting an expanded presence of the sacred in the musical and cultural field, an intention that is confirmed when the author concludes the song using a play on words to dismiss "who is not (re)concave and cannot be (re)convex" (quem não é recôncavo e nem pode ser reconvexo).

The presence of the sacred (axé) in music affected me through another expression used by Camilo Gan, the need to recognise the dimension of the holy to "open the drum" (abrir o tambor), a term that I did not know and that at that moment I believe I understood in a way that may differ from the meaning used there. In the speeches of composers and musicians, I constantly come across reports that there is a moment to be pursued in which music is produced autonomously and in a dimension beyond the rationalisation of execution. I've also heard a similar report from composers for whom the inspiration needed for creation comes from another dimension. Only afterwards would it need to be polished and worked on from the musical craft. Even in the amateur practice of music, I can imagine what the musicians are talking about in these trance moments in which the music and the instrument "open" to those who perform and listen, and how this moment becomes not only the primary purpose of the practice and of the study but the most significant point of connection between musicians and between them and their audience. In instrumental music, this is perhaps clearer to be felt. In contrast, in sung music, when this occurs, the connection between sound and the content of the lyrics reaches an effectiveness that, I believe, enhances and consolidates orality in a sense presented above.

II

On the family's record shelf, most of the records were, in this order, by Caetano Veloso, Gilberto Gil, Chico Buarque, Milton Nascimento, Paulinho da Viola, Tom Jobim, João Bosco, Jorge Ben, Maria Bethânia, Gal Costa, Elis Regina, Rita Lee, Djavan, Gonzaguinha, and Vinicius de Moraes. The rest did not differ much from this nucleus, including many samba records (Martinho da Vila, Zeca Pagodinho, Clara Nunes, Alcione, and Bete Carvalho), northeastern composers (occasional albums by Luiz Gonzaga, Alceu Valença, Novos Baianos, Belchior, and Zé

Ramalho), some bolero records, five records by the Beatles, four by Sinatra, different soundtracks and some international singers. One of the aspects that best unites this collection, in addition to the generational focus, is that most of the artists above built their work from a particular articulation between the consolidation of the complexification of popular songs based on the influence of jazz (filtered by the voice and guitar of João Gilberto), the deepening of the dialogue with Brazilian regional elements (including local aspects of religion) and the critical openness to the international music of the time.

Two artists above had the most significant influence on the way I expanded this discography over the years: Caetano Veloso and his overflow into pre-Bossa Nova influences (Dorival Caymmi, Noel Rosa, Cartola, Lupicínio Rodrigues, Assis Valente, and Ataulfo Alves, among many others) and its critical relationship with national and international contemporary music; and Vinícius de Moraes, the poet of my first children's records and who, much later, I found to have articulated all generations of Brazilian composers and who even introduced me to what he called "Afro-Sambas".

This discography does not include any of the Brazilian gospel music I only got to know in the 1990s and whose sound and content still do not arouse my musical interest, even though today, the gospel industry has a segment that produces songs closer to MPB. In my memory, the presence of the Christian religion is more frequent in pre-Bossa Nova music, whose lyrics resorted to religion to reinforce the drama of a broken heart or a social problem. I also remember the frequent mention of religious festivities (Christmas carols and São João festivities) and songs along the lines of "Ave Maria do morro" (Martins, 1942), which described how the hill is "close to heaven", bringing aspects of religion into everyday life, almost always as a background for another theme. Closer to MPB, I remember Tim Maia's curious foray into the universe of new religiosities on his album "Racional" (Maia, 1985). Attracted by the series of books,[3] the artist composed a double album of effective and potent religious content at the interface between North American soul music and a musicality close to my references.

What I find curious about religious themes in these first memoirs is that, even in Tim Maia's potent work, none aroused any interest in religious practice or even referred to questions close to the artists' spirituality. All passages sounded like a poetic resource or, in the rare cases where religion was the song's central theme, a critical view of the artist about religion. Two pieces exemplify this last look: "Procissão" and "Milagres de um povo".

The song Procissão (Gil, 1964) critiques the social role of religious institutions in the interior of Brazil. The lyrics describe the devotees who "live suffering here on Earth waiting for what Jesus promised" (vivem penando aqui na Terra esperando o que Jesus prometeu), the "people who claim to be God" (gente que se arvora a ser Deus) and the realisation that the "my sertão continues to give to God" (meu sertão continua ao Deus dará), concluding that "if there is Jesus in the firmament, here on Earth this has to end" ("se existe Jesus no firmamento, cá na Terra isso tem que se acabar"). Gilberto Gil, who must have been just over 20 years old when he composed the lyrics and was influenced by student movements at the time, throughout his career, is perhaps the artist who most dialogued with the theme of religion, whether problematising its external dimension (as in the 1997 song Guerra Santa) or allowing transcendental aspects to flow, a much more frequent subject in his work (the 1998 song Kaô would be an example).

Milagres de um povo (Veloso, 1985) opens with a sentence that struck me from the first hearing – "who is an atheist and has seen miracles like me" (quem é ateu e viu milagres como eu) producing a double sensation: the possibility of declaring myself, as my idol, an atheist and, at the same time, recognise the importance of the African-based religion of "the black man who saw cruelty in the face and even produced miracles of faith in the far West" (negro que viu a crueldade bem de frente e ainda produziu milagres de fé no extremo ocidente). In this song, composed for a television series called Tendas dos Milagres, there is also a vast religious vocabulary that I would hear hundreds, perhaps thousands of times, in so many other songs, including Ojuobá, Xangô, Obatalá, Mamãe Oxum, Iemanjá, and Iansã- Hey there, Oba. Like Gilberto Gil, religion in the work of Caetano Veloso, especially that of African origin, appears in its external and transcendental dimension, with the impression that in the latter, the aesthetic, artistic, and cultural dimension takes on more apparent contours and religiosity is more often presented concerning these dimensions.

More broadly, I see in MPB an influence of the intellectualised protest song where religion, the opium of the people, is treated from its institutionality and social role, similar to what happens in Procissão. This same intellectual influence contributes to the treatment of some aspects of religion as a specific culture of a particular region of Brazil, something exotic to be revealed and preserved. It is a relationship with religion that could be read as an "outside look" at the theme. At the same time, in addition to the intellectual influence, I see artists much more committed to the subject from an "inside look".

III

The artists of my training often point out Dorival Caymmi as a genius of the race, the best translator in Bahia, and a master of the written word. His proximity to religion, especially Candomblé, is evident in his songs. Gilberto Gil describes, in Buda Nagô song (Gil, 1992), Caymmi as "a Chinese monk born in black Rome, Salvador" (um monge chinês nascido na Roma negra, Salvador). Antônio Risério[4] makes a broad analysis of the modernity of Caymmi's work and its unique relationship with the effervescent post-industrial urban context of the time, always attentive to how "popular music was an integral part of the process of social projection and organisation of the intelligibility of cultural forms with black-African roots in Brazil" (p. 23).

My relationship with Caymmi, whose work first came to me filtered by João Gilberto, was that of music with minimalist lyrics and melodies, which is why they are modern. I remember that when I heard Caymmi's first albums, that first impression did not coincide with that of the singer with a powerful voice and, above all, who sang traditional themes with solid religious content. In summary, the composer of

Marina, Doralice, Samba da Minha Terra, Saudade da Bahia, Rosa Morena, did not seem to be the same as Suíte dos Pescadores, O bem do mar, Lenda do Abaeté, Dois de Fevereiro, O vento, and many others. In these songs, the dialogue with Candomblé is much more evident and studied by the academy. The orality of these songs, carefully crafted by Caymmi, brings elements of samba de roda, and their verses refer to an ancestral dimension. In the readings and interviews on religion, Caymmi was the reference my memory most often accessed.

Also accessed by my memory during the research was the album Afro-Sambas, released in 1966 by the poet Vinicius de Moraes in partnership with the guitarist Baden Powell (Moraes & Powell, 1966a, b). From the first hearing, I was impressed with how the guitar reproduced the timbre and musical structure of the rituals of African-based religions and how Vinicius' lyrical content is profoundly articulated with the mythology of the Orixás. I remember buying the book Mitologia dos Orixás (Prandi, 2000) motivated by this record. Still, I didn't have, at the time, the opportunity to get to know any terreiros or deepen myself in the study or contact with the theme. Today I recognise some verses as rereadings of myths. Conversely, I identify how the poet makes use of the technology of orality learned in religious songs and "samba de roda" to arrive at the synthesis of content, as we can see in Berimbau's piece from 1966 (Moraes & Powell, 1966a, b): "Whoever does not come out from within himself, will die without loving anyone, the money of those who don't give it, is the work of those who don't have it" (Quem de dentro de si não sai, vai morrer sem amar ninguém, o dinheiro de quem não dá, é o trabalho de quem não tem).

As intense as my relationship with the work of Caymmi and Vinicius is the relationship with the work of Maria Bethânia. I owe her a double debt: the first contact with written poetry (it was through Bethânia that I got to Fernando Pessoa, for example) and the connection with the sacred dimension in music, all mediated by the possibility of having seen several live performances, which, in the case of Maria Bethânia, is very different from remote listening. The artist performed several songs whose central theme is the Orixás and recorded an album dedicated exclusively to songs related to Holy Mary called "Cânticos, Preces e Súplicas à Senhora dos Jardins do Céu" (Bethânia, 2003). More than introducing me to the theme through these recordings, Bethânia's posture on stage and in interviews reveals a ritualistic dimension and is devoted to countless forms of ancestry. By observing his way of singing, especially live, I realised how religiosity could guide and structure the artist's real sensitivity and his search for transcendental forms of musical and poetic expression. The singer's deep concentration on stage establishes a specific pact with her audience that goes beyond performance or musical technique.

Milton Nascimento is the last artist I need to bring to this account and who always takes me to a transcendental dimension. Unlike almost everyone mentioned, Milton has less to do with Bahia and a strong connection with Minas Gerais, where I was born and live. Perhaps for this reason, my religious identification through this artist is not directly linked to Candomblé and ends up going through different references. Although the artist has records and songs that deal specifically with religion – for example, the album Missa dos Quilombos (Nascimento, 1982), it is not his work but the distinction of his voice that most directly refers to the sacred. This association of the figure of Milton and his work with the sacred is shared by close people, and it is not difficult to find statements from other musicians. The impression is that the powerful voice emitted without apparent effort and the very clean falsetto allied to an ancestral musicality and, at the same time, inventive when it incorporates contemporary elements; it creates a magical and monumental atmosphere that awakens the sensation of something transcendent in the listeners. Milton's poetry follows this perception when it deals with broad themes with simple vocabulary but with open and comprehensive meanings, similar but different from what I observed in Caymmi.

There are many other examples that I would have liked to bring here: the transcendental dialogue between Gilberto Gil and Jorge Ben on an album called Oxum & Xangô; the new religiosities in artists such as Zé Ramalho; the affiliation of several of the artists with Candomblé and how this deepened the relationship between music and religion in each work; the use of religious language by samba in Rio de Janeiro; the mass phenomenon of axé music; the brand new pieces that mix ancestral rhythms with new technologies; the use of religion as a distinguishing factor in the musical field, among so many others.

On this last point, about how artists (including those cited in this text) use religiosity as a distinguishing factor to impose a "space of possibilities" in the musical field, I remember Belchior contrasting Caetano's phrase "Everything is wonderfully divine" (Veloso, 1968) with the words "Nothing is divine, nothing is wonderful" (Belchior, 1976) Belchior in Alucinação song (Belchior, 1976) counters in an apparent dispute over the musical narrative of the 1970s: "I am not interested in any theory, not even in those eastern things, astral romances. My hallucination is to endure the day to day, and my delirium is the experience with real things" (*Eu não estou interessado em nenhuma teoria, nem nessas coisas do oriente, romances astrais. A minha alucinação é suportar o dia a dia, e meu delírio é a experiência com coisas reais*).

However, the cited examples were enough to illustrate my perception of how religion crossed my musical formation. This operates in at least three dimensions. The first is how religion was presented to me from outside and inside perspectives, content that contributed to a critical and, at the same time, open and tolerant view of the diversity of religious practices. The second is the perception that, for many musicians, contact with religion opens structuring musical keys, preserved as a sacred and ancestral dimension that must be internalised and activated during the moment of artistic creation and performance. The third is the complexity of how the sung word differs from the written and spoken word, expanding the possibility of orality being created and shared.

Finally, I do not doubt that the MPB that influenced me also influenced several generations of Brazilians and how religion is practised in the country, playing a fundamental role in building respect and possibilities for disseminating religions of African origin. But, on the one hand, if we saw the strong presence of Candomblé in a specific generation of composers, on the other hand, a good part of the Pentecostal expansion is accompanied by a gospel music industry that, just as

MPB affected my worldview, affects the way how the faithful develop their world-view, an aspect to be investigated in future forays into the theme. Also, for future research, it remains to investigate how Candomblé practitioners react to how their faith is appropriated and represented in songs and other productions in the cultural and artistic field.

Notes

1 Artist statement given to the magazine "Revista Civilização", n. 7, 1966. Information obtained from the artist's official website: https://www.caetanoveloso.com.br/biografia/.
2 Camilo Gan is a researcher, musician, percussionist, Afro-Brazilian dancer, samba dancer, choreographer, composer, poet, builder of percussion instruments, restorer of accordions and accordions, music educator, and graduate of Instituto de Ensino Superior Izabela Hendrix. He is the founder and creator of the "Samba de Terreiro" projects, the "Bloco Afro Magia Negra", the band "Trinca Ferro", and the afro dance group "Corpo Oralidade dos Encantados" (Source: Almanaque do Samba, available at https://alma-naquedosamba.com.br; it was accessed on 3 June 2022).
3 Universe in Disenchantment is a collection of books on the philosophy of Rational Culture passed on by the medium Manuel Jacinto Coelho those address topics includ-ing cosmology, metaphysics, ecology, linguistics, theology, UFOs, and flying saucers (Source: Wikipedia).
4 Risério, A. (2011). Caymmi: uma utopia de lugar. São Paulo: Perspectiva.

References

Belchior (1976). Alucinação. In *Alucinação*. Santa Monica, CA: PolyGram/Philips. https://pt.wikipedia.org/wiki/Belchior

Bernardes, T. (2022). Balada de Tim Bernardes. In *Mil coisas invisíveis*. Durham, NC: Coala Records/Psychic Hotline LP.

Bethânia, M. (2003). *Cânticos, Preces e Súplicas à Senhora dos Jardins do Céu*. Rio de Janeiro: Biscoito Fino LP.

Gil, G. (1964). Procissão. In *Louvação*. Santa Monica, CA: Polygram/Philips LP.

Gil, G. (1992). Buda-Nagô. In *Parabolicamará*. WEA LP.

Maia, T. (1985). *Racional*. Vol. 1. Seroma LP.

Martins, H. (1942). Ave Maria do Morro. In *Ave Maria do Morro*. Odeon. 78 RPM.

Moraes, V., & Powell, B. (1966a). *Afro-Sambas*. Vasteras: Forma LP.

Moraes, V., & Powell, B. (1966b). Berimbau. In *Afro-Sambas*. Vasteras: Forma LP.

Nascimento, M. (1982). *Missa dos Quilombos*. Munich: Ariolla LP.

Prandi, R. (2000). *Mitologia dos Orixás*. São Paulo: Companhia das Letras.

Veloso, C. (1968). Divino Maravilhoso. In *Gal Costa (Gal Costa)*. Amsterdam: Philips LP.

Veloso, C. (1985). Milagres de um povo. In *Tenda dos Milagres*. Rio de Janeiro: Som Livre LP.

Veloso, C. (1989). Reconvexo. In *Memória da Pele (Maria Bethânia)*. Santa Monica, CA: Polygram/Philips LP.

8   Beyond space (lessons from the pandemic)

Covid-19 is more than just a biological fact; some sort of 'energy' seems to shake the earth. I often search for information on Instagram and read what my friends share with me. I'm aware that the government is not doing enough, and unfortunately, many people do not respect isolation guidelines. As a result, the number of cases has increased after reopening. In times like these, having faith is extremely important. It's a way to embrace something to guide you and give you the strength to keep going. Looking at it from a biblical perspective offers a bit of explanation about what may be happening. Since science is not providing all the answers, we search within ourselves and find comfort in faith because the crisis is not just biological; there are other forces at play.

(A 23-year-old female, her spiritual beliefs are inspired by Spiritism, Umbanda and astrology. Interviewed in October 2021)

Long shadow of the Catholic Church

Before we delve into the ways in which these religious traditions are represented on the internet, let us first examine their paths in the physical world.

Despite Brazil's formal separation of religion and state, the Catholic Church indelibly impacted the nation's religious landscape, shaping the new context of pluralism and religious freedom. The Church's role in this process can be seen in various aspects, including elaborating the legal framework that regulates the rights and duties of religious groups. The very concept of "religion" in the early years of the Brazilian Republic was heavily influenced by Catholicism, forcing other religious traditions to undergo a gruelling journey towards recognition and legitimacy. The Penal Code of 1890, in its article 157,[1] approved just one year after the establishment of the Republic, criminalised practices deemed as "quackery" and "magic", often associated with religions outside of Catholicism (Giumbelli, 2008). While in the 20th-century Brazil the right to religious expression was legally guaranteed, many other religious traditions, such as Spiritism, Protestantism, Umbanda, and Candomblé, faced discrimination and oppression. The Catholic Church's long shadow loomed large, shaping the religious landscape and influencing the experiences of non-Catholic religious groups in Brazil.

The persecution suffered by Umbanda and Candomblé was not restricted only to criticism of their religious practices and beliefs. It also included the space

DOI: 10.4324/9781003248019-8

characterising a fundamental element of their religious identity, the terreiros. This scenario of repression makes many believers to be discreet about their affiliation, making it difficult to apprehend their exact numbers (Prandi, 2004).

Spiritism faced widespread stigma and persecution due to its healing practices, often seen as fraudulent by the state and other authoritative bodies (Giumbelli, 2008). However, the Protestant and Pentecostal movements have gained prestige and legitimacy, as evidenced by their increasing influence in the media and political arena. Despite making inroads into Brazilian society, Candomblé remains largely marginalised.

Digital Pentecostalism

The rise of Pentecostalism in Brazil has been accompanied by a proliferation of telecommunication networks, including radio stations and TV channels. A survey published in 2017 (ANCINE, 2016) found that between 2012 and 2016, there was a marked increase in the exhibition of religious content on major networks in São Paulo, representing 21% of total programming time. The amount of religious content on national broadcast TV is believed to be even higher, but this cannot be verified due to methodological constraints. Most of this content is produced by Pentecostal churches, particularly the Igreja Universal do Reino de Deus (IURD). As the lines between politics and religion continue to blur and the influence of tele-evangelism grows, we find ourselves bombarded by a deluge of images and narratives that seep into every crevice of our daily lives. Facebook feeds, WhatsApp groups, city streets – it matters not the source, for the result is always the same: a programmed indifference, a numbing to the world around us, a failure to see and truly engage with our surroundings. Simmel's (1950) insights into the blasé attitude of city-dwellers, a naturalised response to the overwhelming stimuli of modern life, ring as true now as they did over a century ago. Yet, in our time's artistic and cultural productions, we see a glimmer of hope, a relativisation of indifference, and a reminder of the power of human connection. From literature to music to Hollywood, the art of our time calls out to us, urging us to look up from our screens and engage with the world in all its messy, beautiful complexity. Audiovisual productions – whether theatre, television, or film – have a remarkable power in shaping our social world. By offering a window into different realities and contexts, they have the ability to foster empathy and understanding, even as they remain grounded in the lived experiences of those around them. The impact of these productions extends far beyond mere entertainment, serving as a crucial complement to life in the bustling cities we call home. It is through these productions that we can reflect upon ourselves, our cities, and the world at large, influencing our habits and customs and shaping the discourses that come to define our collective experience. Take, for instance, the ubiquitous television soap opera, a genre that has come to dominate the airwaves in Brazil. From religious productions to more secular fare, these programmes offer a way to connect with emotions and experiences that might otherwise remain foreign to our daily lives. In this sense, audiovisual production becomes more than mere entertainment; it becomes a tool for self-discovery and a means of exploring the world beyond our immediate surroundings.[2]

As part of our research, we have interviewed several young members of the Universal Church of the Kingdom of God, located in the bustling neighbourhoods of Lourdes and Olaria. These individuals offered a glimpse into the lives of the devout, as they were before the onset of the COVID-19 pandemic and the ensuing social isolation. They were regular attendees at their church, visiting at least four times a week. Their experiences, shaped by the global health crisis, offer valuable insights into the ways in which this particular religious community has been impacted by the pandemic. For the members of IURD, the act of temple-going is a fundamental aspect of their routine, occupying a hallowed place in their schedules and social interactions. The temple serves as a sanctuary, a communal space where individuals of like minds, beliefs, and worldviews can come together, united by the threads of their faith. But what happens when external forces, such as the social isolation policies brought on by the COVID-19 pandemic, threaten to disrupt this delicate balance? How does the church, intricately woven into the fabric of its members' lives, respond to the threat of being silenced? As it turned out, under Jair Bolsonaro, the government came to aid the religious (Christian) communities, recognising the essential nature of their gatherings and allowing for the uninterrupted continuation of their face-to-face activities. But this is just one facet of the church's multifaceted role in offering opportunities for sociability among its members. When asked about the designation of churches and temples as "essential activities", interviewee Emily[3] said, "Because we have a commitment right, the routine of always being going ... I always go only on Sundays because on weekdays I cannot go. Therefore, that is the day I have to go and seek God right". Some interviewees were part of groups such as the Universal Youth Force[4] (FJU), and those that attended the Catedral da Fé church were part of groups on "Whatsapp" to socialise with other church members. During the interviews, we found out that the members of the churches to which the interviewees belong are organised into "families". One could say that there is a consolidation of support and development network within the church, as well as weekly incentives for participants to invite people to join the group, especially young people, to be part of the FJU, which meets every Saturday after the service. The interviewees tried to invite us to attend the groups in person, even during the pandemic. In such a way, the need for physical presence is noted for the spirituality and sociability of groups, even though there are technological tools that allow interactions. As seen in previous chapters, the divine presence is "felt" even outside the temple. Emily said, "It's because, like this, we go to church ... how do I explain it to you. I can have my intimacy at home; anywhere I am, I can have my intimacy with God. I can praise Him (...). In the church, we will be in communion and learning the Word from the Pastor, who understands the Word". Religion, it seems, is inescapable even in the digital realm. The interviews conducted by Pataro, Lima and Arruda in 2021 revealed a recurring theme among young people: their media consumption related to faith or religion. For these individuals, religion is not just a matter of Sunday mornings at the temple, but a ubiquitous presence in their daily lives, permeating the screens of their internet-connected devices, radio, and TV.

As Emily, one of the interviewees, succinctly put it, "I consume media about re-ligion, be it documentaries, music, or worship on the internet. I find myself watch-ing a lot of preaching on Youtube these days". When asked why she turns to online sources for religious content, Emily responded, "We can choose, through the title, the preaching we want to have a deeper understanding of". The primacy of reli-gion in the lives of these young people, and its pervasiveness in the media they consume, is a testament to its enduring influence of IURD in contemporary Brazil.

Beyond the digital

The IURD, it seems, recognises the importance of media in shaping the beliefs and behaviours of the young. According to Leite (2016), the organisation has invested significantly in producing media content that promotes the "constitution of the mo-nogamous, heterosexual family, guided by intelligent faith" to attract new follow-ers and retain existing ones. But the IURD's influence extends beyond the screens of televisions and computers. The organisation also employs "workers", individu-als selected by pastors or approved as volunteers, to serve as spiritual leaders and advisors to the youth. These workers, who perform blessings and spiritual cleans-ing during worship, act as de facto social workers or psychologists, offering guid-ance on personal issues, relationships, and drug use. Their interactions with young people strengthen the bonds between the individual, the group, and the institution. These "workers" serve as gatekeepers, reinforcing the young person's relationship with the IURD early on and demonstrating the organisation's commitment to shap-ing the beliefs and lifestyles of the next generation.

The importance of community and reference to the idea of "family" is a recur-ring theme in the religious experiences of the faithful. Take George,[5] a member of the Candomblé religion, for example. According to him, coexistence, exchange, and life in a community are integral to religious experiences. In Candomblé, "fam-ily" meetings serve as a dogma, as the knowledge of the religion is transmitted orally, and there are no sacred texts. Tradition, culture, teachings, values, and pro-cedures rely on the members' interactions, dialogues, and presence. So it was no surprise that the strict social distancing measures of the quarantine profoundly im-pacted the practice of faith for George and his fellow Candomblé members. The absence of physical contact was deeply felt, as it is a cornerstone of their religious beliefs. But what was unexpected was the resilience and adaptability of the faithful in the face of these restrictions. We found that members of the religious groups we interviewed had other forms of socialisation during the pandemic, such as partici-pating in WhatsApp groups, that allowed them to maintain their bonds even when physical contact was limited. The social cohesion promoted by religion extends beyond the physical walls of the church, transcending the boundaries of time and space to create networks of support and connection through online tools. This is especially true for Pentecostalism, where technology has amplified these religious communities' reach and influence.

The digital realm has proven to be a powerful tool for pentecostalism, allowing its members to organise themselves and reach a wider audience for evangelisation. As Silva (2011) has noted, the use of technology has the potential to circumvent the challenge of attracting people to the church but not retaining them as long-term members. Through social media, pentecostalism maintains social and religious bonds and constitutes a strategy for recruitment.

The principle of individual freedom has also played a key role in the success of pentecostalism in Brazil, as Mariano (2013) has noted. This has led to a mutual relaxation of control between individuals and communities, allowing for greater autonomy and reduced authority over other religious groups. This can be seen in the "intense and growing religious transit in the country" (Mariano, 2013, p. 128). The IURD, in particular, has a set of techniques and strategies designed to attract a large number of people, from the construction of prominent temples and the extensive use of media resources to the provision of a wide range of services (Mariano, 2013). However, those who are attracted to the church in an impersonal manner, such as through flyers and media content or the desire to have a specific issue addressed, may not have strong bonds with other churchgoers and may remain anonymous "in the crowd" (Mariano, 2013). Despite this, Pentecostal pastors and leaders emphasise the importance of collective participation in church meetings and activities, promoting different forms of associative involvement.

Movement and flux

Tele-evangelism is a powerful tool for religious institutions, offering both a means of communication and a platform for proselytism that has the potential to reach a vast audience. The IURD was one of Brazil's first to utilise this resource, but it has since been adopted by numerous other churches. While this may result in a large contingent of sporadic attendees who are in need of further incentives to become long-term members (Mariano, 2013), the floating clientele still provides a significant source of revenue and visibility for the church through their payments for services.

A study in Rio de Janeiro (Fernades, 1998) analysed data from 40,172 households, identified 4,787 residences with evangelical believers, and interviewed 1,500 individuals from this group. The study's results revealed several key characteristics of the sample, including:

- Evangelicals have a high participation rate in church activities, with 85% reporting that they attend weekly, compared to 18% of Catholics and even lower rates for Umbanda and Candomblé.
- The study also found that evangelicals circulate among various denominations at a rate of approximately 25% of the total membership.

These findings reinforce the understanding of religious institutions as infrastructure, providing both a source of spiritual guidance and a certain level of stability (for members) and revenue and visibility (for the institution).

The survey's findings may seem dated, but the scope of its results suggests that the behaviour it describes has not only persisted but has intensified in the decades since its publication. This is evidenced by the significant growth in the number of evangelicals and denominations in the country, as indicated by various analyses of the 2010 Census data. This religious movement is not limited to denominations or branches of a single religion. The establishment of spaces for interlocution between Candomblé and Pentecostalism, for example, leads to symbolic and social exchanges that can result in conversions from Candomblé to Pentecostalism and the modification of religious traditions (Birman, 1996). This was exemplified in one of our interviews with George, who told us of a space for interlocution between the two religions in which his mother, a Pentecostal, watches online masses while the caretaker's wife at his Candomblé house is running online activities. Such examples illustrate the dynamic and evolving nature of religious institutions and the influence of technology on the practices and beliefs of the faithful.

Birman's (1996) research delved into the experiences of former Candomblé and Umbanda members who had joined the IURD and resided in a slum in Rio de Janeiro. They try to keep their affinity with the spiritual realm; however, now they operate within a different symbolic and religious system, utilising their gifts to assist others. The belief held by some Pentecostal groups that Afro-Brazilian religions are inherently evil makes them susceptible to what they perceive as witchcraft and sorcery. As a result, it is not uncommon for pictures of those needing protection to be placed on altars, for anointed oil to be applied to the body, or for cleansing rituals with sulphur to be performed in homes to guard against these perceived threats. While it would be misleading to suggest that this movement of conversion and exchange occurs on a massive scale between numerous terreiros and various Pentecostal denominations in Brazil, it is important to acknowledge that it does exist. Our research has uncovered evidence of this exchange, a type of interaction that is not necessarily based on antagonism but is obviously not without its challenges. During our research, we spoke with a Mãe-de-Santo (leader of a Condomble terreiro), who informed us that she had been consulted by several evangelical pastors. The nature of these consultations was not disclosed, but she mentioned that one pastor only visited her at night and insisted on the utmost discretion. Pentecostal groups approach their religious beliefs with a sense of urgency in their fight against evil, using all the available tools to perform miracles and exercise spiritual gifts.

Despite its lack of social prestige and limited telecommunications infrastructure compared to Pentecostalism, the presence of followers of Afro-Brazilian religions, such as Candomblé, has increased with the democratisation of internet access in Brazil. However, given its history of persecution and relatively small following, one might expect its presence in the digital media landscape to be more reserved. These observations highlight the diverse and sometimes contrasting approaches to religion and the role of technology in shaping the expression and dissemination of religious beliefs and practices.

Ricardo Oliveira de Freitas (2014) argues that this statement is only partially accurate, as the use of the internet by Candomblé followers was primarily limited to correspondence between members for the purpose of organising meetings and discussions about Afro-Brazilian culture. The relationship between media and Afro-Brazilian religions, particularly Candomblé and its emphasis on aesthetic elements, is not recent. Despite being represented in Brazilian popular music and other cultural forms, it has done little to counteract its marginalisation in Brazilian society.

Candomblé followers have embraced the internet in various ways, from promoting social events and festivities to showcasing restricted rituals and ceremonies. Many terreiros, including some considered traditional, have established online presence through websites, Facebook, YouTube, etc. However, there is no universal consensus regarding the use of photography and videography within the terreiros. Certain rituals, such as those involving the incorporation of spirits, cannot be recorded.

Of particular interest to Freitas was the role of "gossip" among Candomblé members in the virtual realm. He suggests that sharing secrets is the foundation of this religious tradition's engagement with digital media. His analysis sheds new light on the appropriation and utilisation of communication technologies within the context of Candomblé.

In 2017, a tale of digital treachery and religious controversy played out in Bahia. The private details of Iarolixá's biological and spiritual family were disclosed on WhatsApp, causing a stir that even the local media couldn't ignore. But as it turns out, such privacy breaches are not as rare as one might assume, and they tend to ripple beyond their original borders, both regionally and globally (Freitas, 2019).

In some ways, the utilisation of digital media by the spiritual community can be seen as an extension of their pre-existing practices and social interactions. The transition to the digital realm only serves to integrate their communicational strategies and modes of self-expression. This has a direct impact on the way they are perceived by society, as the production of "autochthonous content" allows them to write their own narratives and assert their presence in previously unfathomable ways. However, this duality between tradition and innovation raises questions among followers and researchers of Candomblé. On one hand, using modern tools like the internet is crucial for surviving and disseminating knowledge about religion. On the other hand, there are concerns about the potential for the unnecessary disclosure of sacred secrets. This balance between preservation and progress remains a point of contention within the Candomblé community.

The use of digital media by the spiritual community, though perhaps "naive", has the potential to fuel a sensationalising of various aspects of Candomblé's practices and beliefs. To the uninitiated, gossip might seem like nothing more than idle chatter. But, in actuality, it serves a religious purpose, much like the "nagô mail",[6] and is part of a larger context of social inclusion and the assertion of communication rights. The production of content about Candomblé by its members and followers is expanding at an unprecedented rate. People are turning to images and accounts produced by those who attend popular religious events rather than relying on third-party sources. According to Freitas, the public airing of personal details that blur the lines between the supernatural and the mundane is a testament to the distinction between the sacred and the profane is meaningless to the spiritual community. This blurring of reality highlights the interconnectedness of all things, both divine and earthly. Candomblé is a religion steeped in secrecy, where the transmission of advanced knowledge and initiation occurs within the walls of the terreiros, hidden from the public eye. In this context, the relationship between orality and secrecy becomes paramount (Silva, 2011). In Candomblé, the spoken word holds great power, rooted in the belief that it possesses the "force of axé",[7] whereas the written word has less power. During religious rituals, speech transforms from announcements and warnings to those present, conversations with entities, and chants and prayers. The power of speech is tied not only to what is said but also to the authority and legitimacy of the speaker. Over the years, the secrets of Candomblé have been disseminated in various forms, whether through books, music, or the internet, by initiated followers, leaders, and regulars alike. Freitas' research highlights the differing perspectives on the boundaries between public and private within the Candomblé community. These boundaries "extend beyond the confines of their houses, both sacred and secular, to encompass the community as a whole" (Freitas, 2019). Social media and the internet bring followers, terreiros, and society closer together, marking a significant shift from the previous mode of communication that was almost exclusively intra-group.

Additional comments

Being a black Evangelical and a black Candomblé member carries vastly different connotations regarding the social status and positioning. The label "macumbei-ros" is often used pejoratively to refer to Candomblé practitioners, and accusations of using magic persist today, albeit with differing implications for each religious tradition. In Candomblé and Umbanda, magic is a fundamental aspect of their be-liefs, and practices and traditions. Practitioners, often known as "mães-de-santo" and "pais-de-santo", offer magical services and solutions to those in need, making it a legitimate side-professional activity. However, certain Brazilian Pentecostal denominations have characterised this form of magic as malefic, malignant, and evil-oriented, dubbing it "black magic" or "macumba". Books such as "Mãe-de-Santo" written by North-American pastor McAlister (1983) and "Orixás, caboclos & guias. Gods or demons?" by Edir Macedo (1996) have played a significant role in demonising Afro-Brazilian religions. These works argue that Orixás represent evil forces that can only be interacted with through sorcery and witchcraft (Silva, 2007). Despite these efforts to discredit Afro-Brazilian religions, the reaction of its practitioners and supporters has been unwavering and unrelenting. The "Magia Negra" group is one example of resistance, bringing together the Afro-Brazilian community's social, ethical, cultural, and religious dimensions. During our investi-gation, we had the opportunity to meet with one of the group's creators, who shared with us some of their activities and proposals. One of the key goals of this group is to bring together individuals dedicated to combating ethnically and racially moti-vated prejudice towards black people. They aim to spread and celebrate the values of Afro-Brazilian culture through ancestral drumbeats, dance, carnival, and educa-tional activities that "afrobetize" traditional practices. The entry of Candomblé into the digital realm has been instrumental in its pursuit of greater visibility and the ability to assert itself rather than simply being spoken for by intermediaries such as researchers and activists. However, it's worth noting that academic productions and historical and contemporary accounts from travellers, researchers, and missionaries are still sought after and utilised by Candomblé members and even play a role in shaping new practices and interpretations. It put additional responsibility on us and was highlighted by many Candomblé members, who asked us about our contribution to their cause and what kind of feedback our research would yield.

The dynamic evolution of Candomble are partly due to the fact that Candomblé is not a homogeneous religious field, with disputes and disagreements between terreiros from different nations, lineages, and cosmovisions. This has created some obstacles to the flow of information within the community. However, this scenario is starting to change due to democratisation, greater access to the internet, and the artistic, cultural, and academic production of Candomblé followers and practitioners. They are no longer content to simply serve as a data source for research or satisfy the curiosity of those seeking information. We met with Candomblé practitioners who claimed to consume a lot of music and video content related to Candomblé and noticed an increase in content volume during the pandemic, especially regarding topics related to faith. As discussed in the previous chapters, the democratisation of Candomble, partly caused by social media and the internet, goes against the hierarchy of cognitive modes in this religion. If the period of pandemic isolation made Pentecostalism stronger, we would argue that the consequences of this period for followers of the Afro-Brazilian religions are much more complex and not necessarily positive.

Notes

1 Art. 157. "To practice spiritism, magic and its sorceries, to use talismans and cartomancy to arouse feelings of hate or love, to inculcate cures for curable or incurable diseases, in short, to fascinate and subjugate the public credulity: Penalties - of cellular arrest for one to six months and a fine of 100$to 500$000".
2 We have proposed the idea of an "infrastructure of knowledge" to describe how various material and mediated realities construct worldviews of members of IURD (Arruda, Freitas, Lima, Nawratek, & Pataro, 2022).
3 Black female, 23 years old, married, Lives in the municipality of Santa Luzia, metropolitan region of Belo Horizonte.
4 The FJU belongs to the Universal Church of the Kingdom of God (IURD), which aims to reach young people, through actions related to sport, culture and leisure, as a preventive and protective activity in social aspects related to violence, the precariousness of social relations and use or drug trafficking.
5 Twenty-five years old, white male, lawyer.
6 The "nagô mail", also known as "ejó", "fuxico", and "indaka", among others names, is a gossip that was previously disseminated primarily orally by the saint people and nowadays carries its dynamics of information circulation to digital, broadcasting, and printed media (Freitas, 2019).
7 Axé is known in Candomblé as vital force or dynamic force, it is a sacred energy. It is present in everything (Silva, 2011). Popularly, the word axé also indicates a way of greeting, wishing good luck and even a music genre.

References

ANCINE. Agência Nacional do Cinema: Ministério da Cultura. (2016). Informe anual: TV aberta 2016. https://www.gov.br/ancine/pt-br/oca/televisao/arquivos/informe-anual-tv-aberta-2016.pdf

Birman, P. (1996). Cultos de possessão e Pentecostalismo no Brasil: Passagens. *Religião e Sociedade*, *17*, 90–109.

Cavalcanti de Arruda, G. A. A., de Freitas, D. M., Soares Lima, C. M., Nawratek, K., & Miranda Pataro, B. (2022). The production of knowledge through religious and social media infrastructure: World making practices among Brazilian Pentecostals. *Popular Communication*, *20*(3), 208–221.

Fernandes. Sanches, P., Velho, O. G., Carneiro, L. P., Mariz, C., ... Mafra, C., (1998). *Novo nascimento: Os evangélicos em casa, na igreja e na política (Born-again: The Evangelicals at home, in the Church, and in politics)*. Rio de Janeiro: Mauad.

Freitas, R. O. de. (2014). Adeptos do Candomble. In *ECAS 2013 5th European Conference on African Studies African Dynamics in a Multipolar World*. Centro de Estudos Internacionais do Instituto Universitário de Lisboa (ISCTE-IUL), ISBN 978-989-732-364-5.

Freitas, R. O. de. (2019). Candomblé e Internet: ejó, conflito e publicização do privado em mídias digitais. *Tabuleiro De Letras*, *13*(2), 158–172. https://doi.org/10.35499/tl.v13i2.7767

Giumbelli, E. (2008). A presença do religioso no espaço público: modalidades no Brasil. *Religião & Sociedade*, *28*(2), 80–101. https://doi.org/10.1590/s0100-85872008000200005

Leite, M. S. T. (2016). *Força Jovem Universal: estratégias para a juventude da IURD* (Master's thesis). http://www.repositorio-bc.unirio.br:8080/xmlui/bitstream/handle/unirio/11761/Disserta%C3%A7%C3%A3o%20Monique%20S%C3%A11.pdf

Macedo, Edir. (1996) [1988].*Orixás, caboclos e guias: deuses ou demônios?* Rio de Janeiro: Editora Universal.

Mariano, R. (2013). MUDANÇAS NO CAMPO RELIGIOSO BRASILEIRO NO CENSO 2010. *Debates Do NER*, *2*(24), 119–137. https://doi.org/10.22456/1982-8136.43696

McAlister, Robert. (1983) [1968]. *Mãe-de-santo*. 4ª.ed. Rio de Janeiro: Editora Carisma.

Pataro, B. M, Lima, C. M. S. & Arruda, G. A. A. C. (2021). Socialização neopentecostal entre jovens durante a pandemia da COVID-19. *Anais do VIII Simpósio Internacional sobre a Juventude Brasileira, Jubra, Eixo Temático*, 07, 1057–1064.

Prandi, R. (2004). O Brasil com Axé: candomblé e umbanda no mercado religioso. *Estudos avançados*, *18*(52), 223–238.

Silva, V. G. (2011). A questão do segredo no candomblé. Revista de História. Retrieved from https://antropologia.fflch.usp.br/files/u127/segredo.pdf

Silva, V. G. da. (2007). Neopentecostalismo e religiões afro-brasileiras: significados do ataque aos símbolos da herança religiosa africana no Brasil contemporâneo. *Mana: Estudos e Antropologia Social*, *13*(1), 207–236. Retrieved 25 January, 2023, from http://www.scielo.br/pdf/mana/v13n1/a08v13n1.pdf. https://doi.org/10.1590/s0104-93132007000100008

Simmel, G. (1950). *The metropolis and mental life* (pp. 409–424). Adapted by D. Weinstein from Kurt Wolff (Trans.). The sociology of Georg Simmel. New York: Free Press.

9 Conclusions

Demolishing the Babel

Whenever someone enters through the door and says, "I want to start" or "I want to begin," it is believed that the orixá brought that person to us. We have this concept that no one can enter without the permission of the Exu, the Exu at the door is called Onan, and he is known as the Exu of the door. Therefore, when someone enters, we know they have been accepted by Exu. People who come to Candomblé usually face some kind of problem, such as an illness or lack of something. Candomblé is believed to be the solution, and it is referred to as our home, our second home.

(Kelly, Black Candomblecista, interviewed in February 2023)

Undertaking the task of writing this book posed numerous challenges for us as researchers, mainly because we aimed to adopt a "phenomenological-hermeneutical-decolonial" approach. This framework constantly reminded us that we were not discussing our own experiences. In line with Merleau-Ponty and Smith's *Phenomenology of Perception* (1962), we recognised that perception is always the result of bodily experiences and that we were presenting the perceptions of others while registering them through our own experiences as researchers.

Our aim in this book is to help understand the dynamics of urban areas in Brazil. We acknowledge that religion can affect space perception and praxis, and this influence varies across believers of different faiths. It is a fragment of what we have learned through interactions with people and the city over the last four years. To ensure a more participatory construction of our research, we contacted some of the people we interviewed and engaged with professional colleagues to discuss our findings and their potential impact in other fields.

In this chapter, we discuss these dialogues with the people interviewed throughout the research, putting the main arguments outlined throughout the book under discussion. We structured our summary by elaborating on the ideas of each chapter, using the question: "how do the interviewees react to what we are saying concerning ...". This was a semi-structured guide for our discussions, which were triggered unevenly depending on the interviewee's profile. Our interest was twofold: to return part of the knowledge shared with these people and understand their thoughts about our arguments. We strive to ensure that we are not creating a narrative that does not relate to the reality of their experiences. However, the discussions only partly addressed our arguments, instead tended to open new paths of inquiry and

DOI: 10.4324/9781003248019-9

ask further questions. It made us realise that our research will not end, even if this book is finished.

In January and February 2023, we contacted the people interviewed remotely, particularly in the second part of the research. This was necessary as we could not reach the 2018 respondents living in the urban occupation areas during the first stage of the fieldwork. Most people contacted here were interviewed remotely in 2020 and 2021. We also reached out to people interviewed in the final phase of the research in 2022, especially religious leaders living in the Concórdia neighbourhood.

One of the significant challenges we faced while writing the book was embracing interviewees' religious perspectives while remaining respectful and objective as outsiders. We also recognised the importance of a more open and experimental approach to language, gender, and religious views in future research. During the final discussions, we realised that the way how we framed our project and our interest in mapping universal aspects of human behaviours influenced how some interviewees saw religion and what they focused on in the interviews. The most insightful parts of our findings came not as a direct answer to our questions but as long and meandering stories the interviewees wanted us to listen to.

In our project, we recognised the importance of centring the perspectives of our interviewees and Brazilian academics. However, we also acknowledged that this alone would not legitimise our contributions. As (primarily) Brazilian scholars, we understood that we could not simply assume the authority to speak about the experiences and realities of other Brazilians, even within the context of religious studies. The limitations of academic language further compounded this issue, as it often fails to capture the full complexity and nuances of the concepts and categories we seek to explore. This realisation was underscored by Camilo's poignant statement, which pointed out the danger of reducing black culture to a mere spectacle or tourist attraction. He challenged us to delve deeper, to go beyond surface-level observations and engage with the richness and depth of the subject matter. His words reminded us that we must move beyond passive observation and commit ourselves to active engagement and transformation. At the same time, we know we cannot do it while researching other religious groups.

> I'm naturalising this thing about black culture. I'm not saying you all did this, like a safari where you go and get delighted? No! Let's give it more depth because then people are more connected too. There's content for that. We can no longer accept you as observers, especially after this second meeting. You're no longer visitors, even if you have students who didn't come then. It's another class, but you guys are the leaders, right? You already have to commit to activating another future, look, this happened, this and that, it's like this and that over there.

During our discussions with leaders of the Candomble community in Concordia, we learned that the creation of the houses of Candomble is linked to the idea of community, with a strong emphasis on bringing people together to learn from their relatives and pass on this knowledge to their children. Even those not from the blood family can join the community through Candomble and initiation into the religious "family". Thus, Concordia's notion of community is two-dimensional, encompassing the neighbourhood and religious affiliations. The tradition of calling upon others to work together to achieve a common goal, such as building a house or preparing for a festival, is common in many neighbourhoods and communities throughout Brazil. This tradition is rooted in the idea of a social contract, where temporary contracts are formed to achieve a common goal, and once the goal is accomplished, the contract is dissolved. This tradition has a particular temporality and fragility; the bonds between people are negotiated in scope and scale, and the whole mechanism of "communing" is close to what Abdumaliq Simone described in his seminal article "People as infrastructure" (2004).

The Pentecostals and Candomblecistas exhibit this sense of community sentiment but differ in their approach. In Pentecostalism, religion is used as a criterion to bring help to those in need, and individuals are asked about their relationship with God when requesting assistance. Thiago (researcher at UFMG) discussed with us the dual role of Pentecostalism, sometimes in social support and mobilisation networks, sometimes in legitimising neoliberal rationality and individualism in the peripheries. He mentioned that during the distribution of food baskets, it is religiosity that guides the control of who has or does not have access, according to the role of each individual within the relationship networks formed there.

The book explores the concept of sacredness in different places and spaces. Our interviews with various individuals found a complex hierarchy of sacredness in different locations, with some areas being more sacred than others. When we started working on the book, we assumed nothing was profane, but investigating these two religious groups revealed that the meaning of sacredness varies significantly between them. For Candomblecistas, everything can be sacred, and Orixas can be found in any place or activity. There is no real profane space in their worldview. However, for Pentecostals, there is a substantial distinction between sacred and secular space. The secular is not seen as neutral but rather as potentially evil. In the context of this distinction, we had an exciting conversation with Deborah (Kardecist):

> Okay, let's take it step by step. So, what is your conception of the sacred and profane according to my lived experiences and spiritualist beliefs? Before, I had a very Christian conception of the sacred and profane, in the sense of worldly experience, where the profane is associated with the mundane and the sacred is associated with heaven or some external dimension of being. Yeah, and then there's the issue of sacred spaces, right? Like a church, which is a sacred place. When I struggled, I sometimes went to a Catholic church to pray because I felt I would have more contact with God there. Because God is everything. Yes, God is everything. So if God is everything, He is duality too. He is good, and He is bad, you know? Because He is everything, He exists in all dimensions, and in that sense, no space is more sacred than another because your body is sacred, your home is sacred, the air you breathe is sacred, nature is sacred, and the temple is sacred. The profane doesn't exist.

One interesting point of discussion in this chapter was the relationship between religion and the environment. We noted that Pentecostalism has a detached understanding of Nature, viewing it as sacred only if humans leave it untouched. Candomblecistas, on the other hand, do not have a strong distinction between nature and technology. However, none of the explored religions provided a clear framework for dealing with the current climate crisis. We acknowledge that this is a complex and multi-faceted issue not limited to religion. Social and cultural factors influence how we understand and interact with the environment, which shapes religious beliefs and practices. We noted that different Christian segments have varying views on climate change, with some promoting sustainable economic practices and others prioritising economic development at the expense of the environment.

Furthermore, our research has highlighted the limitations of using universal categories and language to understand complex phenomena. The concepts of nature, home, temple, and street are not fixed or universal but are shaped by cultural and contextual factors. Thus, we must approach these categories cautiously and consider their meanings in specific contexts. While the epistemic modes proposed by David G. Robertson may help understand how specific groups act, more is needed to grasp what they think and how they conceptualise the experienced world. The academic tendency to generalise and the will to create useful categories can often lead to a failure to understand what we are looking at. Categories are profoundly contextual and have different meanings for different people in different situations. Therefore, it is essential to acknowledge these differences rather than ignore them to maintain categories as operational elements of academic thinking and writing.

The other thing that caught our attention was the resistance to the use of "theory" to engage with members of the Concordia community, as one could say that they favour another form of knowledge production and dissemination. The use of different epistemic capitals became evident. Even though this approach helped us understand certain categories and concepts operationalised by them, it remains difficult to understand what they actually represent for them. As Kelly commented:

I always say, Little Africa is here in Concórdia, which for me is a teaching, it's a book, right? It comes with the practical part; you can come and do your samba there, singing "2 + 2 is 4" in their heads it's better to understand. This comes from the African roots, it really comes from Africa, because this is really black people's thing, they used to sing to be able to say "tomorrow morning I'm going to open the gate and he or she will come out." Everyone in white starts clapping, thinking it's samba, but we were already there communicating, and that's how we've been maintaining this and bringing it to our activities.

The interactions with the neighbourhood and all its human, spiritual, and social components, the experience of practical living, constitute a specific way of interpreting and navigating empirical reality. We were told that we were taking a class about the culture, history, and religious traditions of the black people as a diaspora and as residents of the state of Minas Gerais that manifest little Africa within the

Concordia neighbourhood. This was summed up for us in one word: "Afrobetiza-tion", a mixture of the words "alphabetisation" and "African".

It is interesting to reflect on the power of music in both Pentecostalism and Candomble and how it has helped shape Brazil's cultural and religious landscape. Music is not only a means of expression and communication but also a powerful tool that can connect people and communities. By recognising the importance of music in these religions, we gain a deeper understanding of Brazil's complex and multi-faceted nature of religious practices.

This book's discussions around language and cultural categories also show how different groups use the internet to organise themselves. For example, Candomble practitioners use internet connectivity to publish activities, while Pentecostals are more open and democratic in sharing their ideas and values. We found an intrigu-ing tension between traditional and younger members of Candomble. The potential consequence of the democratisation of Candomble may threaten tradition and the desire to keep it secret. Candomble is a cultural phenomenon represented every-where in Brazil, but it still remains secretive and hierarchical as a religion.

In the book, we discussed the inward-looking and outward-looking aspects of analysed religions; the expansion of Pentecostalism can be explained, in part, by the desire to sacralise everything and avoid leaving any potentially evil space un-touched. In contrast, the hierarchy of terreiro in Candomble creates a mechanism of distinction, making it difficult for outsiders to understand the concept of the sacred beyond the hierarchy. Candomblecistas prefer to keep their traditions and practices secret, whereas Pentecostals are more open and want to share their faith with oth-ers. When discussing this issue with Lethicia (Baptist Church), she said:

This is not a discussion that is very present in my daily contact with reli-gion, but, more broadly, I can identify three groups: one that believes that everything can attract believers, one that, on the contrary, nothing can, and a group that seeks to find a balance between the principles of religion and the opportunity to expand and spread the word. In my church, I did not experi-ence situations of concern about losing followers or strategies used to gain followers.

Kelly (Candoblecista) approached this issue differently:

I'll be very honest. Oh, our heritage. Our people don't want to, you know, our people don't want to follow. They say, "I want to live my life." Even knowing that they will live, they will fall inside, come back, come back, you know? As I mentioned to Camilo the other day, white people's faith is greater than black people's (...). People don't want that; they don't want to have that patience, so faith is weakening. I told Camilo this; I have several clients ... Most of my clients are white; is it because of their purchasing power? No, it's because of faith (...). It's because their faith is greater, with this idea, that we say, "look, they have their religion over there. Catholic, whatever, we have ours, which is Candomblé; let's separate." So for a long time, they stayed over there, and we stayed over here; in their minds, it was the same thing, but times are changing. It's starting to change, right? That production of whites with blacks, whites are learning to deal with blacks, to have more patience, how to approach them, because it's complicated, right? It's not easy. So I think their faith is a little more than that of blacks. This is not to say that Candomblé is a white religion. But the segment nowadays, the faith, the strength is more towards them.

Camillo clarified the Candomble strategy: "There's no point in fighting with the Western world or just lamenting. We must be inside because that's the only way we can expand regarding knowledge and possibilities".

The growth of Pentecostalism in Brazil since the 1960s has been the subject of much research, with many scholars adopting a functionalist perspective. According to this view, Pentecostalism served an important social function by responding to the anomie caused by the modernisation process and forging socialisation ties that provided alternatives to the traditional social order of country life (Mariano, 2011). The functional analysis emphasised the negative consequences of modernisation and identified them as determinants of Pentecostal demands and agendas. However, other studies have taken a different approach, focusing on the importance of the separation of church and state in Brazil and the breaking of the Catholic Church's religious monopoly. The highly religious nature of Brazilian society created a context of greater pluralism that favoured the emergence of new religious movements. Interestingly, Pentecostalism is viewed as a modernising force promoting the capitalist model and a conservative force fighting against modernising and progressive agendas. This has raised questions about its impact on the urban environment of Belo Horizonte, where its expansion has been significant.

Despite being a planned city, Belo Horizonte's urban expansion has exceeded its established limits, resulting in areas of high population density and peripheries where people from all over Brazil and various religious and cultural expressions are present. Many interviews were conducted for this research in urban occupation areas where the Brazilian State has been absent in providing basic services such as sanitation, drinking water, and electricity. There is a clear predominance of Pentecostal followers in these areas, consistent with many previous studies on the subject. However, we had to consider whether there was a religious component behind this Pentecostal expansion and specifically whether the Pentecostal perception of what is sacred and profane influenced where or the direction of this expansion in the urban context of Belo Horizonte. Our initial investigations sought to answer this question: How does religion affect people's spatial perception in the city? Are there places to be avoided because they are considered profane? Or the opposite, are there privileged places because they are considered sacred? The answers presented in this book are complex and nuanced. We analysed elements of spatial and material religious infrastructure but also subtle individual spatial decisions based on religious worldviews.

The key lesson we learned from exploring different faiths is the paramount importance of context. Though we employed academic categories to compare and understand the diverse experiences of believers, these rigid categories became an insurmountable barrier to genuinely understanding those we conversed with. The orality of religious practices imbues words with a potent meaning that is intimately tied to a specific context, time, and place, inextricably bound to the individuals using them. While conversing with people, we gained a deep understanding of this fact, but when we turned to writing, we found ourselves slowly losing this understanding.

We never aimed to construct a new framework of definitions or categories, and this book is simply a wander through the Brazilian religious landscape, providing readers with guidance and maps. We offer no grand model of the world or universal theory of Brazilian religiosity. We hope that the time spent with us has not been in vain and that now it is time to return home, to the temple, to the street, and engage with nature. We must realise that none of these categories ultimately matter, and this is a good thing.

The Babel, that ancient mythological structure, is often touted as a symbol of humanity's unity. Yet, when we scrutinise the tale more closely, we see it as a machine of oppression, a tool for enforcing a singular, unitarian language upon all. The destruction of Babel was not a punishment for humanity's arrogance but rather a gift of liberation, granting us the freedom to engage creatively in a never-ending, conversational process of understanding the world. For too long, we have been taught to worship at the altar of linguistic uniformity, bowing down to a monolithic conception of language and thought. But the truth is that the diversity of languages celebrates humanity's boundless imagination and endless capacity for expression. So let us abandon the false idols of linguistic uniformity and embrace the glorious polyphony of tongues surrounding us, for it is in this diversity that we find human communication's true richness and beauty.

References

Mariano, R. (2011). Sociologia do crescimento pentecostal no Brasil: um balanço. *Perspectiva Teológica, 43*(119), 11. https://doi.org/10.20911/21768757v43n119p11/2011

Merleau-Ponty, M., & Smith, C. (1962). *Phenomenology of perception* (Vol. 2012). London: Routledge.

Simone, A. (2004). People as infrastructure: Intersecting fragments in Johannesburg. *Public Culture, 16*(3), 407–429.

Index

Printed in the United States
by Baker & Taylor Publisher Services